Macmillan
ENCYCLOPEDIA OF SCIENCE

9

Fuel and Power

Clint Twist

Macmillan Publishing Company
New York

Maxwell Macmillan International Publishing Group
New York   Oxford   Singapore   Sydney

Published by:
Macmillan Publishing Company
A Division of Macmillan, Inc.
866 Third Avenue, New York, NY 10022

Collier Macmillan Canada, Inc.
1200 Eglinton Avenue East, Suite 200
Don Mills, Ontario M3C 3N1

Planned and produced by Andromeda Oxford Ltd.

Library of Congress Cataloging-in-Publication Data

Macmillan encyclopedia of science.
        p.   cm.
   Includes bibliographical references and index.
   Summary: An encyclopedia of science and technology, covering
such areas as the Earth, the ocean, plants and animals, medicine,
agriculture, manufacturing, and transportation.
   ISBN 0-02-941346-X (set)
   1. Science–Encyclopedias, Juvenile. 2. Engineering–
Encyclopedias, Juvenile. 3. Technology–Encyclopedias, Juvenile.
[1. Science–Encyclopedias. 2. Technology–Encyclopedias.]
I. Macmillan Publishing Company   90-19940
Q121.M27   1991   CIP
503 – dc20   AC

Volumes of the *Macmillan Encyclopedia of Science*
 1 *Matter and Energy* ISBN 0-02-941141-6
 2 *The Heavens* ISBN 0-02-941142-4
 3 *The Earth* ISBN 0-02-941143-2
 4 *Life on Earth* ISBN 0-02-941144-0
 5 *Plants and Animals* ISBN 0-02-941145-9
 6 *Body and Health* ISBN 0-02-941146-7
 7 *The Environment* ISBN 0-02-941147-5
 8 *Industry* ISBN 0-02-941341-9
 9 *Fuel and Power* ISBN 0-02-941342-7
10 *Transportation* ISBN 0-02-941343-5
11 *Communication* ISBN 0-02-941344-3
12 *Tools and Tomorrow* ISBN 0-02-941345-1

Printed in the United States of America

Introduction

Energy and engines keep the wheels of civilization moving. This volume examines energy sources from fossil fuels to nuclear fission and the Sun, describes the generation and use of electricity, and surveys power-producing machines from steam engines to rocket engines.

To learn about a specific topic, start by consulting the Index at the end of the book. You can find references throughout the encyclopedia to the topic by turning to the final Index, covering all 12 volumes, located in Volume 12.

If you come across an unfamiliar word while using this book, the Glossary may be of help. A list of key abbreviations can be found on page 87. If you want to learn more about the subjects covered in the book, the Further Reading section is a good place to begin.

Scientists tend to express measurements in units belonging to the "International System," which incorporates metric units. This encyclopedia accordingly uses metric units (with American equivalents also given in the main text). For detailed information on units of measurement, and on numbers, see pages 86-87.

Contents

Part One

Energy sources

Without fuels and other sources of energy, human existence would be cold, dark, and primitive. The discovery of fuel, in the form of firewood, is as old as the discovery of fire itself; but there is far too little firewood in the world to support modern civilization. At present, most of our energy comes from burning fossil fuels: coal, oil, and gas. Coal is now our basic fuel; there is a plentiful supply and we have enough for hundreds of years. Oil and gas are much more useful, but we have very limited resources, enough perhaps for only another fifty to one hundred years.

In the 1970s, nuclear energy was widely believed to be the energy source of the future. Today, however, there is a large question mark over the nuclear energy industry. Attention has now turned instead to natural sources of power: solar energy, waterpower, wind power, wave power, tidal power, and our planet's own reserves of heat energy.

◀ Solar cells, also called photovoltaic cells, produce small amounts of electricity from the energy in sunlight. Solar cells are most efficient when they are used in outer space.

Coal

Coal is a form of fossilized wood, and coalfields are the remains of great forests that existed hundreds of millions of years ago, during the Carboniferous period. For the last two hundred years, coal has been our most important industrial fuel, and today it supplies the major source of our electricity. In countries where large coalfields occur, coal mining is a major industry that uses some of the world's largest machines.

Coal is our most plentiful fuel, and total reserves are enough to last for hundreds of years. Although coal mining damages the environment, and burning coal causes pollution, it is certain to remain a vital source of energy for the foreseeable future.

▶ Shoveling coal deep underground. Although coal mining is now highly mechanized, it still involves a great deal of human labor in very difficult conditions. In some places, coal miners have to work on their hands and knees in tunnels with a roof height of less than 1.5 meters.

Fields and reserves

Coal is a black or dark-colored mineral fuel that consists mainly of the element carbon. As well as carbon, coal contains a variety of hydrocarbons, often traces of sulfur, and moisture. Coal occurs as seams (layers in the rock strata) within the Earth's crust. The depth at which coal seams are found varies enormously. In some coalfields, the seam lies just below the surface. More often, coal seams are buried beneath hundreds of meters of rock.

The properties of coal have been known for at least 2,000 years, but coal was little used until the Industrial Revolution of the 1700s. Before then, firewood and charcoal provided virtually all the world's energy needs. During the 1700s coal quickly replaced charcoal as an industrial fuel. In the 1800s, coal provided the driving energy behind the great Age of Steam. Today, coal is used mainly to produce electricity.

▼ World coal reserves are estimated at around 10 quadrillion metric tons. Most of them are in the Northern Hemisphere, although there are some important deposits in Australia. The United States, the Soviet Union and China between them have over 60 percent of Earth's total resources.

Energy from coal

Total world coal production stands at over 4 billion metric tons per year. This accounts for just less than one-third (31 percent) of the planet's total energy production. Coal is still widely used in factory and home furnaces, but most of the world's output is burned in power plants to produce electricity. Although it is sometimes considered an old-fashioned fuel, coal is in fact our biggest single source of electrical power, and will long continue to be so.

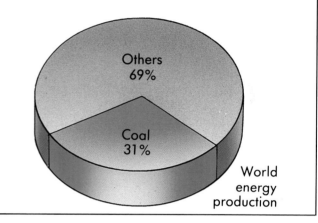

Others 69%

Coal 31%

World energy production

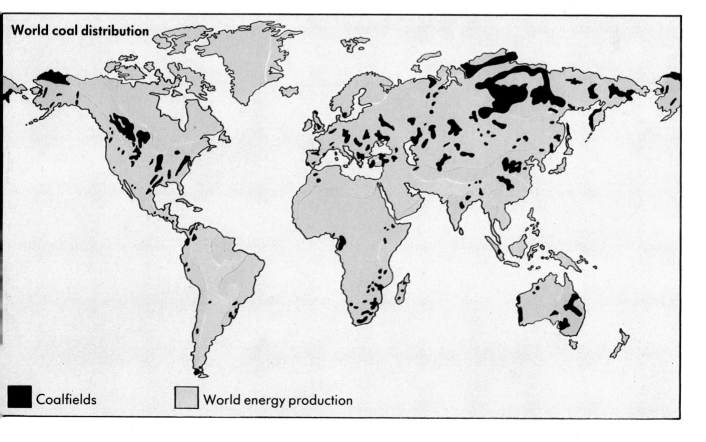

World coal distribution

■ Coalfields   □ World energy production

Coal formation

Coal was formed by the carbonization of trees and plants. When plants die, the carbon in their tissues is normally recycled back into the environment during decomposition. Carbonization occurs when dead plant material is subjected to heat and pressure over millions of years. The different grades of coal were formed by different combinations of time, heat, and pressure.

Most of the world's coal was formed during the Carboniferous period, between 360 and 286 million years ago. At this time, large areas of the Earth's surface were covered with dense, swampy forests. Dead plants and trees that fell into the swamps did not decompose completely, but accumulated into thick layers of wet peat. When the swamps were later flooded by the sea, the peat became buried under layers of sediment. Over long periods of time, it decayed further and slowly dried and hardened into brown coal or lignite.

As further layers of sediment built up, increased heat and pressure caused lignite to turn into bituminous coal. In some instances, additional pressure turned bituminous coal into anthracite.

Anthracite coal is known as hard coal because of its rocklike appearance. Anthracite is the best-quality coal, and contains 86 to 98 percent carbon. It burns with a bright blue flame and gives off very little smoke. Bituminous coal is the most widely occurring grade of coal, and contains less than 86 percent carbon. Bituminous coal also contains the highest proportion of volatile material, which can be distilled into gas and coal tar. One of the most important discoveries of the Industrial Revolution was the process of baking bituminous coal in an oven to produce coke.

Lignite and brown coals are sometimes soft enough to be crumbled between the fingers. They contain less carbon than does bituminous coal, and are often compressed into pellets or briquettes before use.

▼ Fossil remains of the plant *Neuopteris fexuosa* found in coal. The plant lived in the Carboniferous Period (360-286 million years ago) and when the coal-forming material was laid down, it was trapped. Although the plant material decomposed, it left a clear impression in the coal seam.

▲ Coal forms where dead plant material does not properly decay. First it forms peat (1). Once buried and compressed, it becomes lignite (2). Further pressure produces bituminous coal (3) and anthracite (4). The plants were once part of swampy forests in the Carboniferous period (left).

Nearly coal – peat

The water in swamps and bogs often does not contain enough oxygen or bacteria for normal decomposition to take place. Dead plant material slowly forms a layer of waterlogged peat. Over thousands of years, the layers build up and can reach 30 m (100 ft.) in thickness. Although peat is still some way from being coal, it can be used as a low-grade fuel. When it is freshly cut from the ground, peat is a black slimy material containing about 70 percent water. After it has been dried, it is a crumbly, brown solid. When burned, peat gives off large quantities of thick smoke. In many rural parts of Europe, peat is still cut by hand in the traditional manner and is burned for domestic heating. In some countries it is cut by machines and used in small power plants. Most of Europe's peat reserves are now nearly depleted.

Surface mining

When a coal seam occurs just below ground level, it can be worked by surface, or strip, mining. The overburden, the rocks and soil that cover the seam, is removed, and the exposed coal is mined by mechanical excavators.

In the simplest surface, or open-pit, method of mining, the coal is dug mechanically out of an excavation. This is used for relatively thick seams with little overburden. For thinner seams, a widely used technique involves cutting a series of trenches. When all the coal exposed by a particular trench has been removed, another trench is dug alongside and the overburden may be used to fill the previous trench.

In Germany's Ruhr Valley, the coal seam lies very close to the surface, and it is only necessary to scrape aside the topsoil with draglines. In parts of the United States, however, up to 60 m (200 ft.) of overburden lie above the coal seam.

In order to be economic, strip mining needs to be carried out on a very large scale. Some of the power shovels used can remove over 150 cubic meters (200 cu. yd.) of rock or coal with a single bite.

In general, it is the lower grades of coal that lie nearest the surface, and which are recovered by surface mining. This is especially true in the western United States and Eastern Europe where huge deposits of brown coal and lignite are worked. Some high-quality coal is also obtained by strip mining, including almost half the anthracite produced in the United States.

The main problem with surface mining is the damage it creates to the environment, particularly if the coal lies beneath farmland. In many countries, strip mining is now carefully regulated. Some governments require that topsoil be removed and stored separately so that it may be replaced and the area replanted with a minimum of delay.

▲ A giant walking dragline at work in a British coalfield. Draglines of this size require very large and level working surfaces in order to operate efficiently. The ordinary excavator in the foreground of the picture is dwarfed by comparison. It is used to clear obstacles from the dragline's path.

◄ A strip mine in Australia, showing the extensive damage that surface mining can cause to a landscape.

► A power shovel in an American strip mine. Even a very hard coal such as anthracite is soft enough to be worked by mechanical shovels and draglines. Brown coal and lignite are soft enough to be scooped out by bucket-wheel excavators.

Underground mining

▲ Typical hoisting gear at the pithead, often the only visible feature of an underground coal mine. This mine is in the Rhondda Valley in Wales. Early mines were not very deep, and coal was carried to the surface up a series of ladders.

Underground mining is much more widespread than surface mining in some countries. Great Britain, for example, obtains more than 90 percent of its coal from underground, and mines are often more than 1,000 m (3,000 ft.) deep. Access to the coal seams is obtained by digging a vertical shaft. Mechanical hoisting gear raises and lowers cages for miners and coal.

Mining is carried out along a series of horizontal tunnels and branching galleries. The exposed portion of the seam that is being worked is called the coal face. Some narrow seams (less than a meter, or yard) are still worked by hand. Thicker seams are worked by modern coal-cutting machinery that can cut up to 6,000 metric tons of coal per day.

There are two basic underground coal mining techniques: room-and-pillar and longwall. Room-and-pillar mining, which is still widely used in the United States, involves removing coal from a series of underground rooms. Large pillars of coal are left in place to support the roof of the gallery.

Rotary coal cutter

Electric locomotive

Room-and-pillar mining

Hydraulic roof supports

Longwall coal face

Underground coal mine

▲ Modern mines have at least two shafts, one for coal, the other for miners and their equipment. A ventilation system pumps fresh air down one shaft and forces stale air out of the other. Miners may travel up to 2 km (over 1 mi.) by train to the coal face. Longwall mining enables coal to be cut by rotary cutters and be carried along conveyor belts into loading hoppers. At the surface, the coal is sorted and graded according to size and quality.

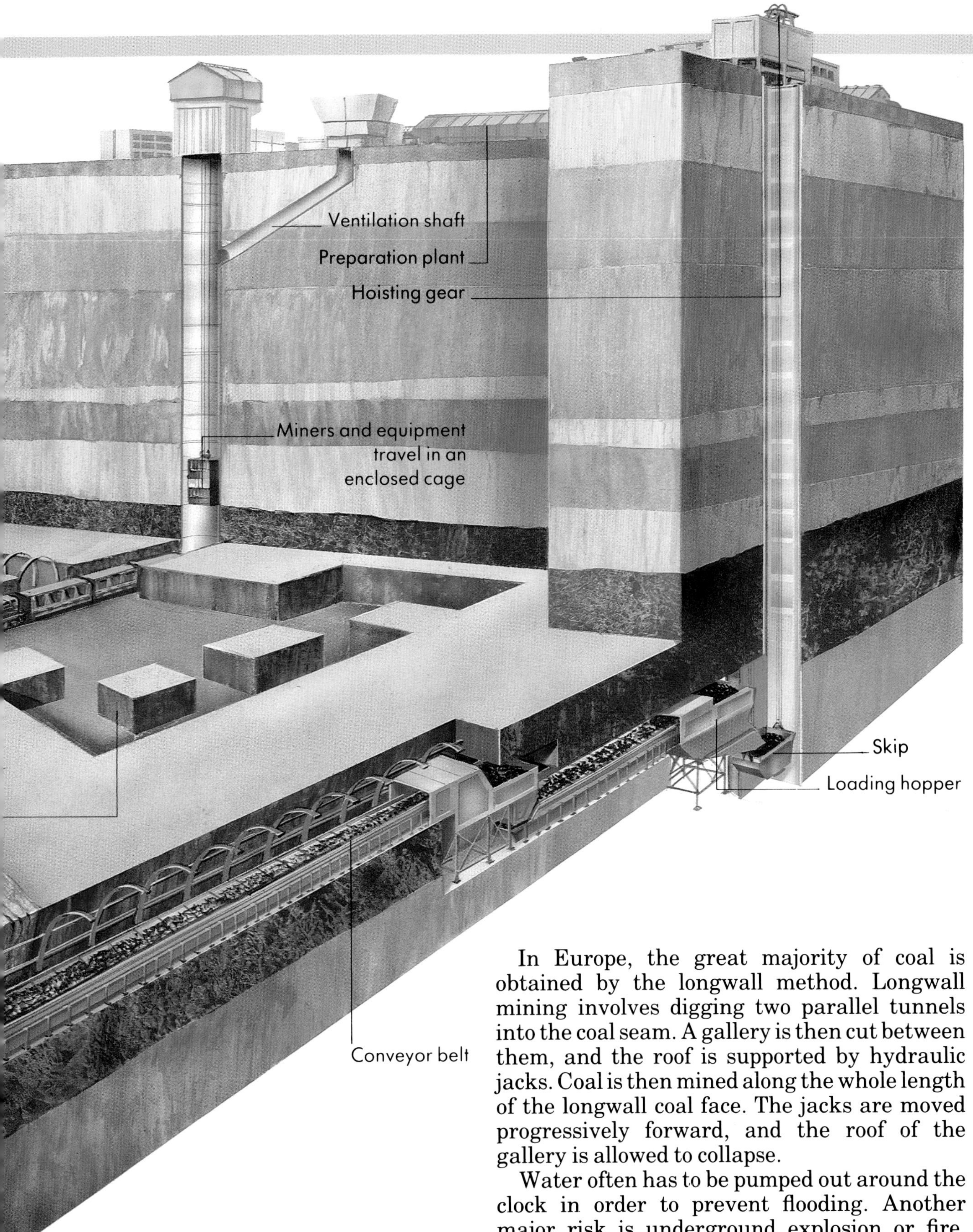

Ventilation shaft

Preparation plant

Hoisting gear

Miners and equipment
travel in an
enclosed cage

Skip

Loading hopper

Conveyor belt

In Europe, the great majority of coal is obtained by the longwall method. Longwall mining involves digging two parallel tunnels into the coal seam. A gallery is then cut between them, and the roof is supported by hydraulic jacks. Coal is then mined along the whole length of the longwall coal face. The jacks are moved progressively forward, and the roof of the gallery is allowed to collapse.

Water often has to be pumped out around the clock in order to prevent flooding. Another major risk is underground explosion or fire, because of natural gas and coal dust suspended in the air. Great care must be taken not to produce sparks, which could ignite coal dust, and electrical equipment is heavily insulated.

Oil and gas

Spot facts

- *Virtually all forms of powered transport now use fuels that can be extracted from crude oil.*

- *The Trans-Alaska Pipeline was designed to carry two million barrels of oil per day.*

- *More than 40,000 oil-fields have so far been discovered, but fewer than 4,000 of them have any commercial importance.*

- *The United States has already used up more than half its original oil reserves.*

- *The world's largest natural gas field is in Siberia, and lies only about 1,200 m (4,000 ft.) below the surface.*

▶ Oil production platforms in the North Sea. Small deposits of natural gas that occur in oil fields are normally burned off at the surface in flares. Larger deposits of gas represent a valuable source of energy in their own right. Natural gas is now the world's third most important source of energy.

Crude oil, or petroleum, is the most valuable and the most versatile of Earth's buried treasures. Oil provides us with a number of different fuels, each of which is essential to modern civilization. Gasoline, aviation fuel, diesel fuel, and heating oil are all refined from crude oil. During the last 100 years, the search for oil has spread to ever more remote and difficult areas: hot deserts, offshore waters, and polar wastes.

More recently, natural gas has also emerged as an important fuel, and in many parts of the world it is piped directly into houses for domestic heating and cooking. Oil and gas are frequently found together.

Fields and reserves

By weight, oil and natural gas consist almost entirely of the elements carbon and hydrogen. Chemically, these two elements are combined into thousands of different compounds known as hydrocarbons. Oil and gas occur in natural underground reservoirs that may lie thousands of meters below the surface.

Both oil and gas have been known since ancient times, but they were very little used before the 1860s. The first oil well was drilled in Pennsylvania in 1859. Within a few decades, oil had also been discovered in other countries. The most recently developed oil fields are in the polar regions of Alaska and Siberia.

Oil has become the world's most important fuel, which, when refined, is used for domestic and industrial heating and for powering the engines of our machines. Natural gas was once considered a waste product, but is now widely exploited as a heating fuel.

▼ Total world oil reserves are estimated at around one trillion barrels. The largest reserves are in the Middle East, which has 26 supergiant fields. Each supergiant field contains at least five billion barrels.

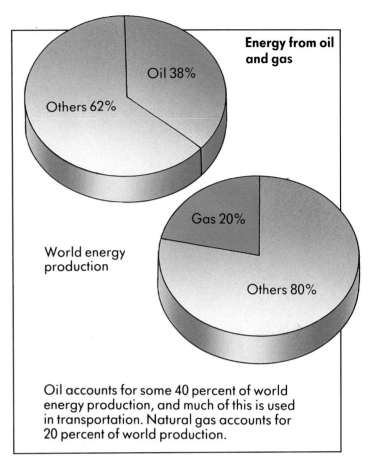

Energy from oil and gas

Oil 38%

Others 62%

World energy production

Gas 20%

Others 80%

Oil accounts for some 40 percent of world energy production, and much of this is used in transportation. Natural gas accounts for 20 percent of world production.

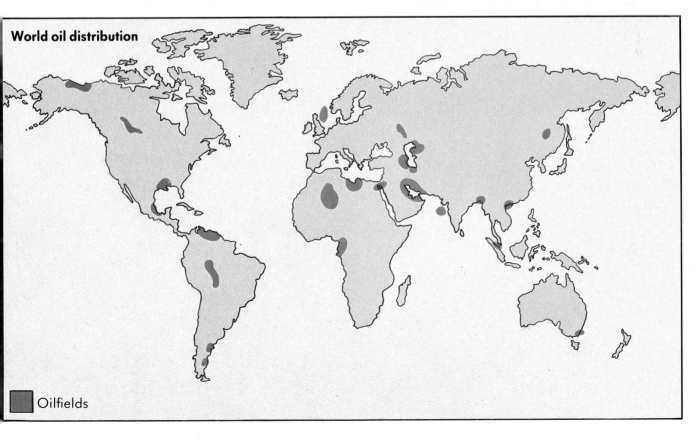

World oil distribution

⬛ Oilfields

Formation

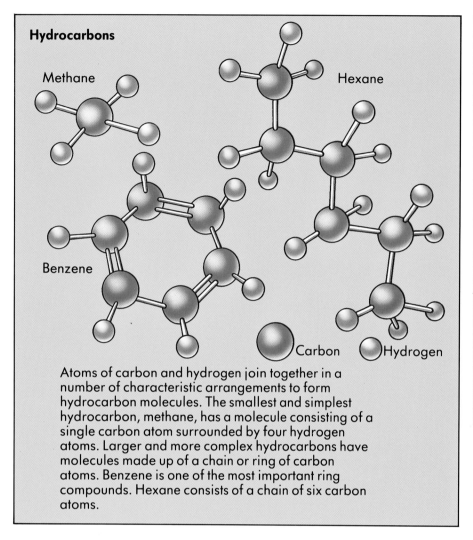

Hydrocarbons

Methane

Hexane

Benzene

Carbon   Hydrogen

Atoms of carbon and hydrogen join together in a number of characteristic arrangements to form hydrocarbon molecules. The smallest and simplest hydrocarbon, methane, has a molecule consisting of a single carbon atom surrounded by four hydrogen atoms. Larger and more complex hydrocarbons have molecules made up of a chain or ring of carbon atoms. Benzene is one of the most important ring compounds. Hexane consists of a chain of six carbon atoms.

▲ Pitch Lake in Trinidad, a naturally occurring deposit that has seeped out on to the surface, forming a lake of liquid bitumen (asphalt). Bitumen is one of the heaviest hydrocarbon compounds.

Oil was formed from the remains of tiny plants and animals that lived in the oceans many millions of years ago. The exact age of oil deposits is hard to determine, because oil moves about easily and is not usually found in the rocks in which it was formed.

At some periods of Earth's geological history, the remains of algae and plankton accumulated on the seabed and were buried under layers of sediment. The sediment preserved the organic matter from the process of decomposition. Instead, it was transformed by the action of bacteria into a substance known as kerogen. Over long periods of time, further layers of sediment produced increased temperature and pressure, which "cooked" the kerogen and produced many different hydrocarbons. Depending on the exact recipe of the "cooking," crude oil can be thick and dark, or pale and thin.

Oil is chemically stable within the Earth's crust, but a number of factors cause it to migrate physically. As a liquid, oil tends to move through the narrow spaces between particles of rock by capillary action. In rock that is saturated with water, this movement is always in an upward direction because oil is less dense than water. Many oil fields are literally floating on underground water.

Like water, oil passes easily through permeable rocks such as sandstone and limestone, but cannot penetrate impermeable rocks such as slate. Some oil deposits migrate all the way to the surface, but most are eventually trapped by a layer of cap rock. The commonest type of oil trap is an anticline, an arch in the rock strata caused by the folding of the Earth's crust. About 80 percent of world oil production comes from anticline deposits.

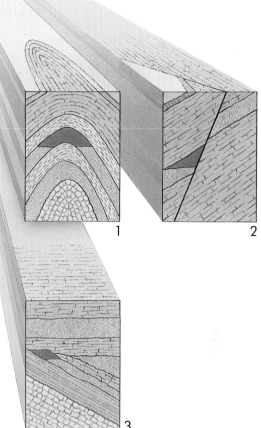

Key
1 Anticline
2 Fault
3 Unconformity
4 Intrusion
5 Pinch-out
6 Salt dome

▲◄ As it migrates upward through permeable rock, oil may become trapped by impermeable cap rock in a number of geological features: (1) an anticline, or dome; (2) a fault caused by Earth movements; (3) an unconformity (an abrupt change in the nature of the rock strata); (4) an intrusion (caused by a mass of rock pushing up from below); (5) a pinch-out (where the rock strata taper away); and (6) a salt dome (caused by the intrusion of a large pillar of rock salt).

Natural gas is often found in association with oil, but also occurs as separate deposits. Most gas was formed by the same processes that created oil, although gas formation can take place over a wider range of geological conditions. Some gas has also been created by freshwater swamps or volcanic activity, and some has leaked from coal deposits. Gas is more mobile than oil, and may therefore migrate along different routes and be trapped in different geological formations.

The best-quality natural gas is dry gas, which consists of almost pure methane, with small amounts of ethane. Natural gas that also contains significant amounts of propane, butane, and pentane is known as wet gas.

Production

Modern surveying techniques can predict the location of an oil field with a fair degree of accuracy. The exact nature and value of the field has then to be proved by drilling a series of test holes to obtain samples. Only then can the location of the production wells be determined. Even on oil fields that are already in production, as many as four out of five wells that are sunk may turn out to be dry.

Under ideal conditions, natural pressure from gas and water trapped with the oil will cause it to flow up the well to the surface. Where there is no natural pressure, or where it has become exhausted, the oil must be pumped to the surface. In some cases, oil can be forced to flow up the well by pumping water into the oil field at a different location. Other techniques, such as underground explosions, have also been used to stimulate flow from unproductive wells.

About five percent of world oil production comes from heavy oil, which is much thicker than other crude oils. In many cases, heavy oil can be extracted only by pumping steam down the well to increase the rate of flow.

The deepest wells bring oil from more than 10,000 m (6 mi.) underground, but many fields lie within the first few thousand meters of rock. Faults and layers of drill-resistant rock are a much greater problem than depth. Such obstacles can be overcome by drilling at an angle, using a wedge-shaped attachment known as a whipstock.

In a few countries, oil is obtained from oil shale, a type of deposit in which oil is physically locked into the rock. Brazil, the Soviet Union, and China produce small quantities of "synthetic" crude oil from oil shale. But at present the process is expensive.

Undersea production

Extracting oil from beneath the seabed requires specialized marine technology to overcome the additional barrier of seawater. Offshore oil production began during the 1920s, but was restricted to simple platforms on stilts built in shallow coastal waters. During the 1970s, the search for oil was extended to the deeper waters of the North Sea and the Gulf of Mexico. Today, oil is produced from wells drilled in up to 300 m (1,000 ft.) of water.

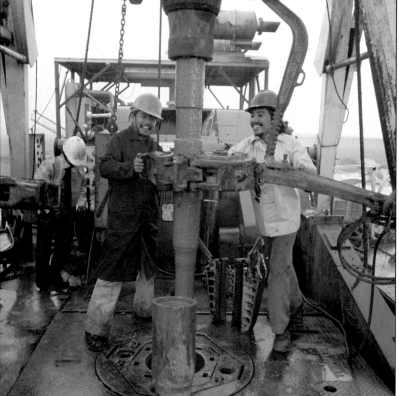

◄ When the oil in a field is under natural pressure, the flow must be regulated by a complicated system of pipes and valves known as a "Christmas tree."

▼ The pumps used to bring up oil are often known as nodding donkeys because of their ceaseless up-and-down motion.

Offshore drilling

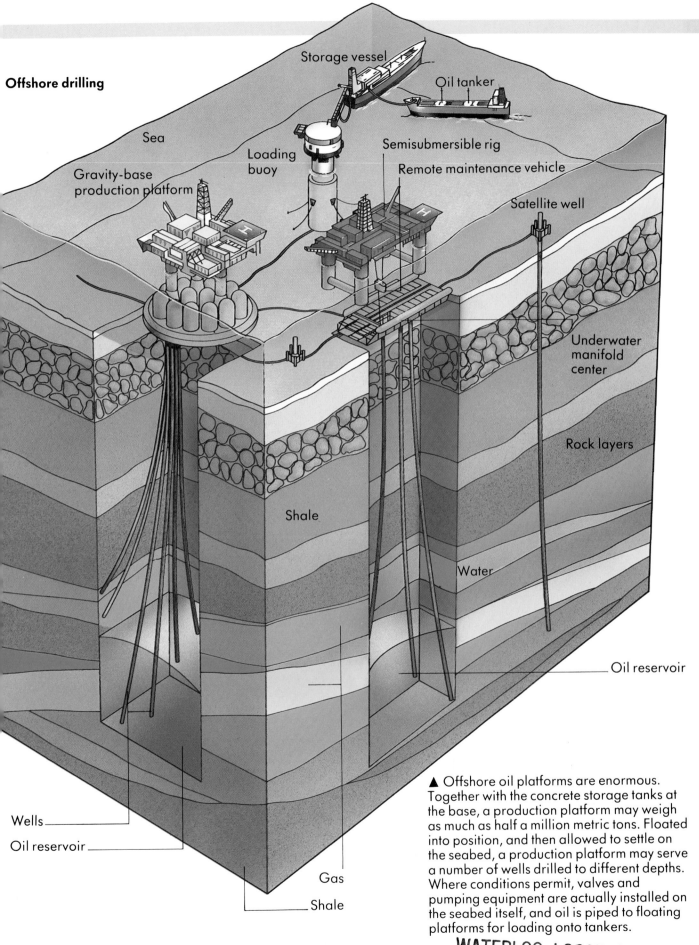

Storage vessel

Oil tanker

Sea

Loading
buoy

Semisubmersible rig

Remote maintenance vehicle

Gravity-base
production platform

Satellite well

Underwater
manifold
center

Rock layers

Shale

Water

Oil reservoir

Wells

Oil reservoir

Gas

Shale

▲ Offshore oil platforms are enormous.
Together with the concrete storage tanks at
the base, a production platform may weigh
as much as half a million metric tons. Floated
into position, and then allowed to settle on
the seabed, a production platform may serve
a number of wells drilled to different depths.
Where conditions permit, valves and
pumping equipment are actually installed on
the seabed itself, and oil is piped to floating
platforms for loading onto tankers.

Transportation

Transporting oil

Oil is constantly on the move, and every day about 50 million barrels are transported somewhere around the globe. The basic unit of the oil industry is the barrel, which contains 159 liters (42 gallons), but actual barrels are very little used today. The scale of the global trade in oil demands transportation in bulk, either by pipeline or in specially designed ships.

The largest ships, known as supertankers or VLCCs (Very Large Crude Carriers) measure over 300 m (1,000 ft.) in length, and can carry up to 3.5 million barrels of oil at a time. The oil is usually loaded and unloaded at specially constructed deep-water ports. In some places, for example the Persian Gulf, oil is first piped to floating platforms anchored some distance offshore. In some offshore oil fields, oil is loaded onto tankers directly from an underwater installation by a process known as wellheading. In other offshore fields, for example in the North Sea, the oil is carried from the production wells by pipelines to storage tanks onshore.

In general, pipelines are used over short distances and for local distribution. The diameter of the pipe varies between 20 and 120 cm, (8 and 48 in.) depending on the volume to be carried. The construction of long-distance pipelines can present tremendous engineering problems. Where these problems can be overcome, however, pipelines offer the cheapest form of bulk transportation.

One of the most demanding pipeline projects ever undertaken was the Trans-Alaska Pipeline. It was built to transport crude oil from Prudhoe Bay in the Arctic, south to the ice-free port of Valdez. During construction, engineers had to brave blizzards and temperatures down to −50°C (−60°F). They had to cross countless rivers, deal with the problem of permafrost, or permanently frozen ground, and allow for possible earthquakes and migrating caribou.

▼ A liquefied petroleum gas tanker unloading its cargo into storage tanks. Natural gas can be stored in underground caves and disused mines.

Transporting gas

Before gas can be transported, its volume must be reduced by compression and cooling. Natural gas can then be carried by high-pressure pipelines, or in liquid form on specially constructed ships. The liquefied petroleum gases (LPGs), such as butane and propane, liquefy fairly easily. They can be stored and transported at normal temperatures. Liquefied natural gas (LNG), which is liquid methane, requires constant refrigeration.

▼ (main picture) A section of the 1,300-km (800-mi) long Trans-Alaska Pipeline snaking across the frozen wastes of Alaska. In the northern part of the state, the pipeline is constructed above ground. This is necessary because the oil passing through is warm and the ground beneath is permanently frozen. Piping the oil through the ground would cause the permafrost to melt, leading to severe environmental damage.

▼ (inset) A team of pipe-laying machines work in concert to lower a section of the Trans-Siberian Pipeline into a trench. This pipeline carries natural gas 6,000 km (nearly 4,000 mi.) from Siberia into Europe.

Refining

Crude oil is of little immediate use, because the individual hydrocarbons are completely mixed together. Useful fuels and other products have to be separated from the crude oil by refining. An oil refinery operates around the clock, processing up to 200,000 barrels of oil per day.

The main refining process is that of distillation, which is based on the fact that different substances vaporize and condense at differing temperatures. Hydrocarbons with small, light molecules, for example methane and butane, are gases at room temperature. The heaviest hydrocarbons, such as asphalt, are almost solid. The liquid fuels – heating oil, diesel, gasoline and kerosene, all vaporize at fairly low temperatures.

Simple distillation releases only a small amount of impure fuel. Refineries use a sophisticated process known as fractional distillation to separate crude oil into its component parts, or fractions. The process is carried out inside a hollow steel tower known as a fractionating column. In large refineries these rise more than 45 m (150 ft.) above the ground, and contain 30-40 separate condensing plates.

After fractionation, the heavy oils that are left can still be made to produce useful fuel by cracking them. When heated under pressure, their molecules split up into lighter, more valuable hydrocarbons. The cracking process is accelerated if the hot oil is passed through catalysts such as natural and artificial clays.

The most modern refineries use molecular sieves made from certain dehydrated minerals. Depending on the final product required, the broken molecules may then be recombined into other compounds by a second catalytic process. Pumping hydrogen into the catalyst also boosts the efficiency of the cracking process.

Other techniques are also employed. Liquid petroleum gases are sometimes extracted by the absorption process, in which hydrogen is bubbled through the crude oil. The gases are then recovered by washing the hydrogen with steam. Solvents and acids are often used at the end of the refining process to remove impurities.

Refineries also produce petroleum spirits, waxes, greases, and carbon black for printing ink. Many other products are used as feedstock (raw material) for the petrochemical industry.

Fractionating column

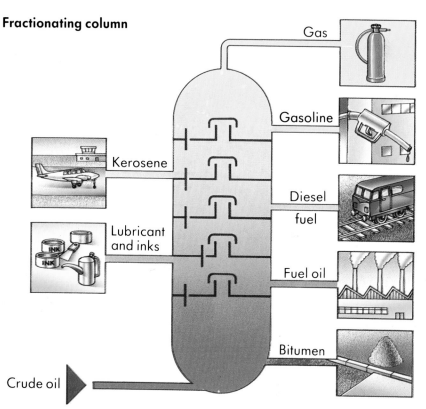

◀ Crude oil is heated to about 350°C (660°F) before it is pumped into the bottom of the column. Oil vapor then rises up the column through a series of steel trays, each of which contains a large number of condensation traps. The trays are maintained at slightly different temperatures, gradually getting cooler toward the top of the column. Different fuels and other products condense in the trays at different heights, and are tapped off. Any gas in the crude oil passes out of the top of the column, and is piped to a gas separation plant.

▶ The lights of an oil refinery glow against the night sky. An oil refinery involves several complex processes. After fractionating, some of the heavier oils are passed to the catalyst cracking plant to be broken into lighter grades. Sulfur may be removed from diesel fuel. The heavier fractions may be further processed, and a vacuum distillation plant produces lubricating oil and paraffin wax.

Nuclear power

▶ The control room at the Calder Hall nuclear power station in Britain. Calder Hall was the first commercial nuclear reactor, and it commenced operation in 1956. Today there are more than 400 nuclear power reactors located in over 25 different countries.

Uranium is the rarest of all naturally occurring "fuels," and it is also the most powerful. Inside a nuclear reactor, uranium can be made to release energy from the very heart of each atom. Most of this energy is in the form of useful heat. Nuclear power plants use only small amounts of uranium fuel, but produce large quantities of electricity.

Unfortunately, uranium is also our most potentially dangerous naturally occurring fuel. As well as heat, uranium emits potentially harmful radiation. Many people believe that nuclear power production presents a serious threat to the planet Earth. In some countries, the subject of nuclear energy is now highly controversial.

Safe energy?

The first experimental nuclear power plant started operating in the United States in 1951. Initially, there was considerable enthusiasm for this new source of energy. Nuclear power promised to supply the world with large quantities of clean, cheap electricity, especially in countries that lacked reserves of coal and oil. By the early 1960s, more than 100 nuclear power stations had been built. Developing nations were especially eager to acquire this new technology. Today, however, a question mark hangs over the nuclear power industry.

The great advantage of nuclear energy is that it uses very little fuel. The great disadvantage of nuclear energy is that it produces high levels of harmful radiation. In 1986, an accident caused an explosion at the Chernobyl nuclear power station in the Soviet Union. The explosion created a cloud of radioactive material that contaminated land and therefore food supplies over large areas of the Soviet Union and Europe.

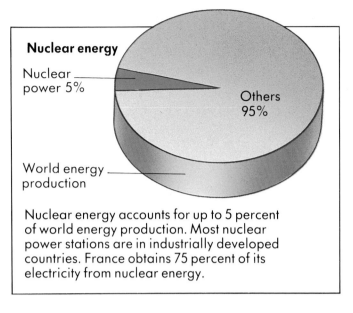

Nuclear energy

Nuclear power 5%

Others 95%

World energy production

Nuclear energy accounts for up to 5 percent of world energy production. Most nuclear power stations are in industrially developed countries. France obtains 75 percent of its electricity from nuclear energy.

▼ The nuclear power station at Three Mile Island, Pennsylvania. In 1979, an accident caused the core of a reactor to overheat so much that it began to melt (inset). If the core had melted down completely, huge amounts of radioactivity could have been released, threatening the lives of thousands.

How it works

Nuclear energy is the energy released during the fission, or splitting, of uranium or plutonium atoms. As well as releasing energy, the fission of, for example, a uranium atom also releases neutrons. Some of these strike other uranium atoms, causing them to split, thus releasing more energy and more neutrons. This process is known as a chain reaction. Uranium is the only naturally occurring element in which a chain reaction can take place.

Once it has started, a chain reaction tends to accelerate until all the uranium is consumed. Under particular circumstances the chain reaction can be made to happen almost instantaneously. This produces the awesome destructive power of an atomic bomb. Huge amounts of energy are released, but it cannot be put to any constructive purpose.

A nuclear reactor is a device for producing a slow, controlled chain reaction. The energy that is produced is released at useful levels over long periods of time.

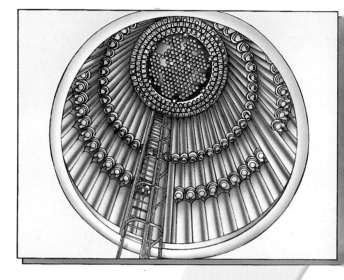

▲ Uranium fuel is loaded into a large number of cylindrical metal containers known as fuel rods. These are then packed closely together to form the core of the reactor. The shape of the core varies with different designs of reactor.

Inside the atom

A typical atom consists of a nucleus surrounded by shells of orbiting electrons. The nucleus is composed of protons and neutrons, held together by an incredibly strong force. When an atom splits, some of this force is converted into energy.

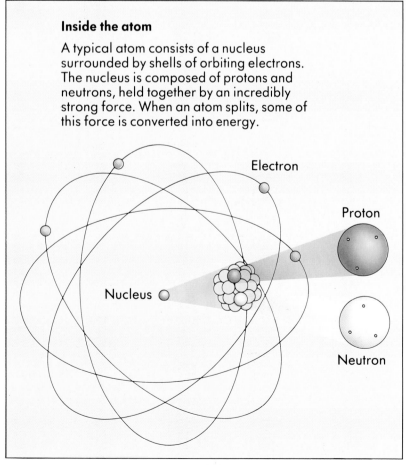

Electron

Proton

Nucleus

Neutron

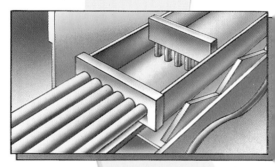

▲ The ore has to be refined in order to produce uranium fuel, which is shaped into thin rods or compressed into pellets.

▼ Mining uranium ore. Even good-quality ore may contain as little as 1-2 percent uranium. The largest deposits are in North America.

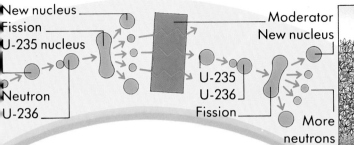

New nucleus
Fission
U-235 nucleus

Neutron
U-236

Moderator
New nucleus

U-235
U-236
Fission

More
neutrons

▲ A chain reaction starts with an atom of, say, uranium being struck by a stray neutron. Fission then produces more neutrons, which enable the chain reaction to proceed. The moderator helps control the speed of the reaction.

▲ The dangers of radioactivity mean that great care must be taken when transporting nuclear fuel. Special railroad cars are often used. These are designed to withstand a crash at speeds of up to 150 km/h (90 mph).

► All forms of radioactive waste, including the clothing worn by workers, must be carefully sealed and stored. Waste with only a small amount of radioactivity is often stored in steel drums.

▼ Using nuclear fuel requires a number of different processes. Uranium must be processed before it can be loaded into a reactor. After use, the fuel must be reprocessed for safe disposal.

◄ Highly radioactive waste must be treated with the utmost care. At this French installation, nuclear waste is being sealed into glass blocks. This reduces the risk of a leakage of radioactivity into the environment.

Reactors

A nuclear reactor has three basic components: a core, a coolant system, and a containment. The core produces heat, and the coolant system carries the heat away from the reactor. Most coolant systems operate under high pressure, and the whole reactor is therefore encased in a strong reactor vessel. The containment is an outer covering, usually made of reinforced concrete, that prevents radiation escaping.

The core consists of the fuel rods arranged within a moderator. The moderator serves to slow down neutrons, because slower neutrons bring about fission more readily. The intensity of the chain reaction can be adjusted by a series of control rods made from substances that absorb neutrons. Lowering the control rods causes the chain reaction to slow down. Raising them speeds it up.

The different types of nuclear reactor are designed to make use of different grades of fuel. Natural uranium metal can be used as a fuel only if it is surrounded by an extremely efficient moderator, such as graphite. The uranium is formed into fuel elements, and as many as 30,000 may be stacked into a graphite core measuring up to 15 m (50 ft.) high. Such reactors built in Great Britain have a coolant system that uses high-pressure carbon dioxide gas.

The majority of reactors currently in use run on uranium fuel which has been improved by the process of enrichment. Instead of uranium metal, they use a compound known as uranium dioxide. Many reactors that run on enriched fuel use ordinary water, both as a moderator and as a coolant.

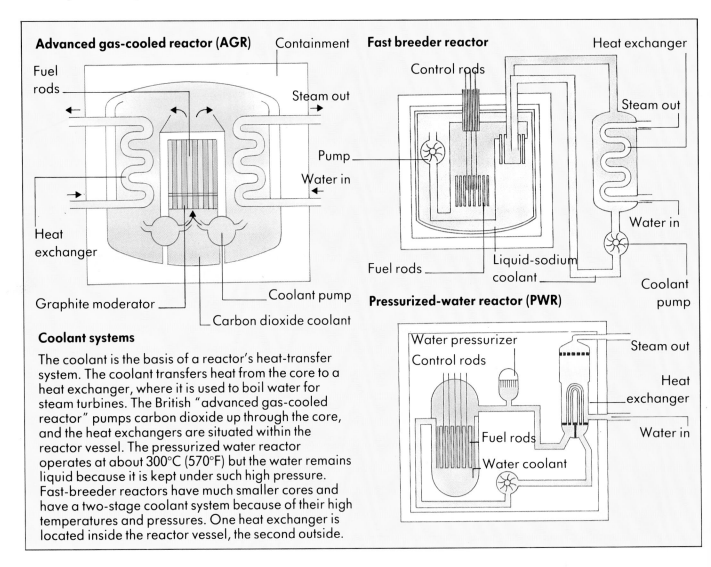

Advanced gas-cooled reactor (AGR)
Containment
Fuel rods
Steam out
Pump
Water in
Heat exchanger
Graphite moderator
Coolant pump
Carbon dioxide coolant

Fast breeder reactor
Heat exchanger
Control rods
Steam out
Water in
Fuel rods
Liquid-sodium coolant
Coolant pump

Pressurized-water reactor (PWR)
Water pressurizer
Control rods
Steam out
Heat exchanger
Water in
Fuel rods
Water coolant

Coolant systems

The coolant is the basis of a reactor's heat-transfer system. The coolant transfers heat from the core to a heat exchanger, where it is used to boil water for steam turbines. The British "advanced gas-cooled reactor" pumps carbon dioxide up through the core, and the heat exchangers are situated within the reactor vessel. The pressurized water reactor operates at about 300°C (570°F) but the water remains liquid because it is kept under such high pressure. Fast-breeder reactors have much smaller cores and have a two-stage coolant system because of their high temperatures and pressures. One heat exchanger is located inside the reactor vessel, the second outside.

Most water-cooled reactors use water at over 100 times atmospheric pressure, and are known as pressurized-water reactors (PWR). Other designs make use of boiling water as a coolant.

If uranium fuel is very highly enriched, it can be used in reactors that do not need a moderator. This type of reactor is known as a fast reactor. Fast reactors can also make use of plutonium, an element extracted from depleted fuel. One major advantage of fast reactors is that they can be used to "breed" more plutonium from depleted uranium fuel. Such reactors are often called breeder reactors.

Fast reactors operate at higher temperatures than other reactor types, and produce energy more efficiently. There are several designs, with liquid sodium the customary coolant.

▼ The chart shows the electricity generating capacity of nuclear power plants ordered by European countries during the period 1956-85. Nuclear power became most popular when oil prices rose during the 1970s. Since the accident at Three Mile Island in 1979, nuclear energy has undergone a worldwide decline in popularity. This is clearly indicated by the number of new reactors ordered in Europe after 1980.

Hands off

Even small doses of radiation can be harmful to human health. Workers at nuclear power stations take every precaution to prevent their exposure to radiation. Inside the reactor building, workers wear heavy protective suits lined with radiation shielding. Delicate operations, such as removing depleted fuel rods, are usually carried out by remote control from behind heavily shielded walls.

Commercial orders for nuclear reactors

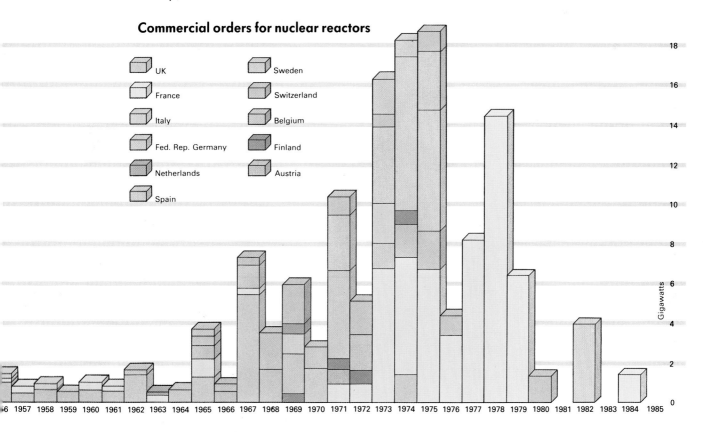

UK
France
Italy
Fed. Rep. Germany
Netherlands
Spain
Sweden
Switzerland
Belgium
Finland
Austria

Gigawatts

Solar energy

▶ Located about 150 million km (93 million miles) away, the Sun is a huge nuclear furnace that radiates vast quantities of energy into space in all directions. Only a very small proportion (about a billionth) of that energy reaches Earth.

The Sun represents an inexhaustible source of free energy. Most buildings already make some use of passive solar heating, and in many countries the Sun's energy is actively collected to provide hot water for household purposes.

The main drawback with solar energy is that it produces only low temperatures under natural conditions. In order to produce useful quantities of electricity from solar energy, the heat energy in sunlight must be collected over a large area and concentrated at a single point. In countries with suitable climates, experimental solar-energy power stations are in operation.

Sun power

Solar energy dwarfs all our other energy sources. In less than one hour, the Earth receives energy from the Sun that is equivalent to the world's total energy output from other sources during an entire year.

Most of the Sun's energy is reflected back into space or is absorbed by the atmosphere. However, sunlight still reaches the Earth's surface in usable quantities. On a summer's day in Michigan, for example, the energy falling on one square meter (about 1 sq. yd.) of sunlit ground is equivalent to ten 100-watt light bulbs.

Sunlight provides a constant source of energy for the Earth as a whole, but it is not evenly distributed over the planet's surface. Solar energy can be exploited only where and when the sun is shining. Usually this means hot countries with clear skies. Even in cold countries, however, solar energy can be very useful.

The energy in sunlight can most easily be exploited in the form of direct heat. Rooms can be heated simply by letting the Sun shine in freely. Higher temperatures, needed to provide domestic hot water, require the active technology of solar collection panels.

Sunlight can also be converted directly into electricity using solar cells. At present, these are mainly used in calculators and watches. But larger solar-powered devices also work. A solar-powered aircraft has flown between Great Britain and France; and solar-powered cars have been built in several countries.

▼ Some 30 percent of the Sun's energy that reaches Earth is reflected back into space by the atmosphere. Virtually all of the remaining 70 percent is absorbed by the atmosphere, where it powers the water cycle. Much of the absorbed heat is reradiated into space. A small part of the solar energy heats the Earth's surface, causing winds and currents.

Solar energy

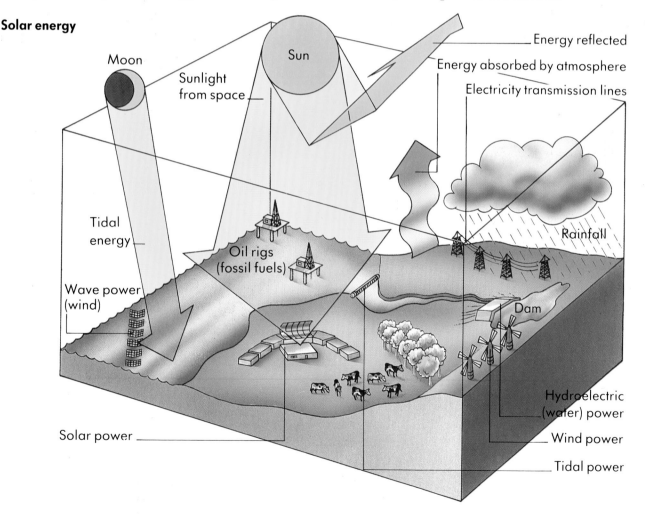

Moon

Sun

Sunlight from space

Energy reflected

Energy absorbed by atmosphere

Electricity transmission lines

Tidal energy

Oil rigs (fossil fuels)

Rainfall

Wave power (wind)

Dam

Solar power

Hydroelectric (water) power

Wind power

Tidal power

At home

All houses and other buildings already make use of solar heating from sunlight shining on walls and rooftops, and through windows. Even in countries which have cool, cloudy climates, houses obtain as much as 20 percent of their space heating (room heating) requirements from the Sun. By incorporating passive solar technology, such as sloping windows and Trombe walls, this can be boosted up to 80 percent. This sort of heating is known as solar gain. Houses that are designed to make maximum use of solar gain are usually well insulated in order to keep heat loss to a minimum.

Producing domestic hot water from the Sun requires a solar collector. The commonest type is the flat-plate collector, which consists of tubes inside a sealed glass-topped box. The plate or tubing is black because dark-colored surfaces absorb heat better than lighter ones. The sealed box reduces heat loss. The water in the tubing may go directly to a hot-water storage tank. Alternatively, the water may flow in a closed system, and heat is transferred through a heat exchanger to the hot-water tank. This is more efficient than using the Sun-warmed water directly from the pipe.

Millions of these rooftop solar collectors are now in use worldwide, particularly in the countries around the Mediterranean Sea.

▼ This British house makes good use of solar gain. Large sloping windows are the simplest form of solar technology. A sloping surface can receive up to 10 percent more solar energy than a vertical one. This principle is used in many solar-energy devices.

▲ This spacious house uses a number of light-sensitive switches to retain the heat obtained from solar gain. When the Sun is shining, the windows open and air can circulate through the house. When the Sun goes behind a cloud, the switches close the windows to trap the heat.

Evacuated-tube collectors can also be used to supply houses with hot water. An evacuated-tube collector may consist of a blackened glass cylinder inside a sealed glass tube. A vacuum inside the tube insulates the inner cylinder against heat loss. Heat is transferred by air or liquid circulating within the inner cylinder. A typical installation may contain 20 or 30 tubes connected together.

Evacuated-tube collectors are more expensive than flat-plate collectors, but can deliver about twice as much useful energy over a year. By surrounding the tubes with curved mirrors, or by focusing the Sun's rays through lenses, higher temperatures and greater efficiency can be achieved.

Active and passive

In "active" solar-heating systems a working fluid, such as water, is pumped around a closed circuit. An example is a flat-plate collector where the Sun's energy is used to heat water and that heat is then transferred to a separate supply of water. A Trombe wall is a form of passive technology. Sunlight shining through the glass outer wall is absorbed by the dark-colored inner wall. Convection currents circulate warm air into the room, and draw cold air out.

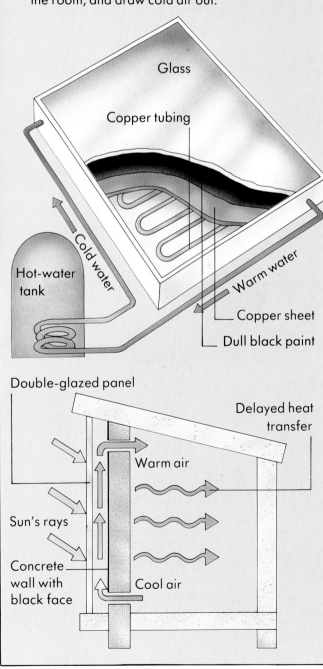

Solar electricity

Producing useful quantities of electricity from solar energy requires a very large-scale installation. The most widely-adopted design is that of the solar field. This is already being used in several countries including Australia, Japan, Spain, Italy, and the United States.

A solar field consists of many rows of individual solar collectors. These are connected to a central heat exchanger that produces steam to drive an electric generator. The collectors are normally surrounded by curved reflectors, and are made even more efficient by the process of Sun tracking. Each collector is mounted so that it can be swiveled and tilted to be always facing directly at the Sun. Throughout the day, the position of the collectors are constantly adjusted by small computer-controlled motors.

The main disadvantage with the solar field is that heat energy is lost during the transfer from the collectors to the central heat exchanger. One solution to this problem is to concentrate the Sun's heat into a central collection point by using a circular field containing thousands of separate mirrors. The first central collection systems were experimental solar furnaces that could reach temperatures above 3,000°C (5,400°F). During the 1980s, however, the first central collection power towers began operating. The Sun's rays are focused at the top of the power tower, and heat is collected by a series of black-colored pipes containing liquid sodium. Heat exchangers at the base of the tower are connected to boilers that produce steam to drive generators.

Photovoltaic cells can also be used to produce large quantities of electricity, but at present the process is too expensive to be practical. They are more efficient in space than on Earth.

▶ The world's largest power tower at Barstow, California. The tower itself stands about 90 m (295 ft.) tall, and the field of mirrors covers some 90,000 square meters (over 100,000 sq. yd.)

Solar cells

A solar cell, often called a photovoltaic (PV) cell, converts the energy in sunlight directly into electricity. An individual cell consists of two thin slices of silicon crystal sandwiched between two layers of metal. The top layer of metal is in the form of a grid so that sunlight can reach the upper side of the silicon. The two slices of silicon contain slightly different amounts of impurities, causing them to have different electrical states. Sunlight falling on the upper slice causes electrons to flow into it from the lower slice. This creates an electrical current that flows through the metal contacts. The photo shows panels of solar cells mounted on a research satellite. In space, the cells will receive the full strength of the Sun's rays, and will be able to operate at maximum efficiency. On the Earth's surface, however, even strong sunlight has had most of the energy filtered from it by the atmosphere.

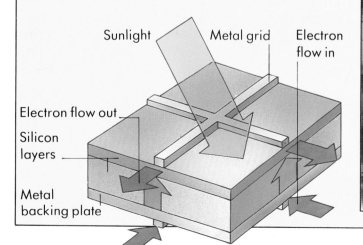

Sunlight Metal grid Electron flow in

Electron flow out

Silicon layers

Metal backing plate

Energy from the elements

▶ Waves crashing ashore are a constant reminder of the energy that the seas and oceans contain. Our coastlines offer tremendous potential for the future, but wave-power devices are still at the scale-model stage. At present, the cost of building full-sized machines is far too high to be practicable.

Our planet is rich in natural sources of energy. Waterpower, wind power, and wave power provide further opportunities to use the energy of the Sun. Solar heat powers the water cycle, which provides rainfall and running water. Uneven heating of the Earth's surface causes the winds to blow, and at sea the wind creates waves. The daily rise and fall of the tides, however, is caused by the gravitational pull of the Moon.

The motion of water, wind, waves, and tides can all be harnessed by machines, with varying degrees of success.

Geothermal energy is the heat energy found in rocks deep below our planet's surface. This source of energy can be used more directly.

Nature's power

Nearly all of nature's power that we harness comes in the form of movement. It is largely a case of converting one sort of movement into another. The movement of running water and the rushing wind is converted into rotation by machines based on the wheel.

A waterwheel is placed edge-on into a flow of water. The wheel is turned by the force of the flow against blades set across the wheel's rim.

In a traditional windmill, the wheel takes the form of a number of angled blades, or sails, which are placed facing the wind. A windmill turns because the wind is deflected by the angled blades as it flows through the wheel.

Waterwheels and windmills have been in widespread use for at least 2,000 years. Since Roman times they have provided useful energy for milling grains for flour, or for pumping water for irrigation and drainage. At the beginning of the Industrial Revolution, waterpower provided most of the energy that drove the spinning wheels and other machinery in the first factories.

During the last 100 years, the energy conversion process has been taken one step further. Waterpower and wind power are now used to drive turbines. These in turn are used to generate energy in the form of electricity.

▶ Traditional windmills provided a steady source of low-speed rotation that was very useful for certain tasks such as grinding grain or pumping water.

▼ An ancient waterwheel in Syria, possibly dating from Roman times. Waterwheels could be almost any size, but until the 1700s the materials used were very weak.

Waterpower

Water is the most useful source of natural power because it is the easiest to control. Running water can be channeled along sluices and through pipes. More importantly, a river can be blocked by a dam, creating a reservoir that can store huge quantities of water. Water from the reservoir can then be released as and when it is required.

Waterpower is harnessed to generate electricity in hydroelectric power plants. These are usually situated at the base of a large dam. The best locations are the narrow, steep-sided river valleys found in mountainous areas. A dam across such a valley can create a reservoir more than 100 km (60 mi.) long. Large-scale projects may involve more than a simple dam and reservoir. In the Snowy Mountains of Australia, the waters of the Snowy River have been diverted by a series of underground tunnels to some 16 hydroelectric plants.

Waterpower can also be used to store surplus energy from other power plants. This is carried out in what are known as pumped-storage plants. These use two separate reservoirs at different levels.

During normal operation, water from the upper reservoir is used to drive turbines to produce electricity. After passing through the turbines, the water is stored in the lower reservoir. Whenever there is a surplus of electricity, it is used to pump water from the lower reservoir back up into the higher one. Demand for electricity is at its highest during the day. This means that, in most stations, pumping is often done at night.

▼ Construction workers inside the tunnels at a hydro plant The worker on the right is in the main water-supply tunnel. The worker on the left is standing at the mouth of an intake pipe leading to a turbine.

Hydroelectricity

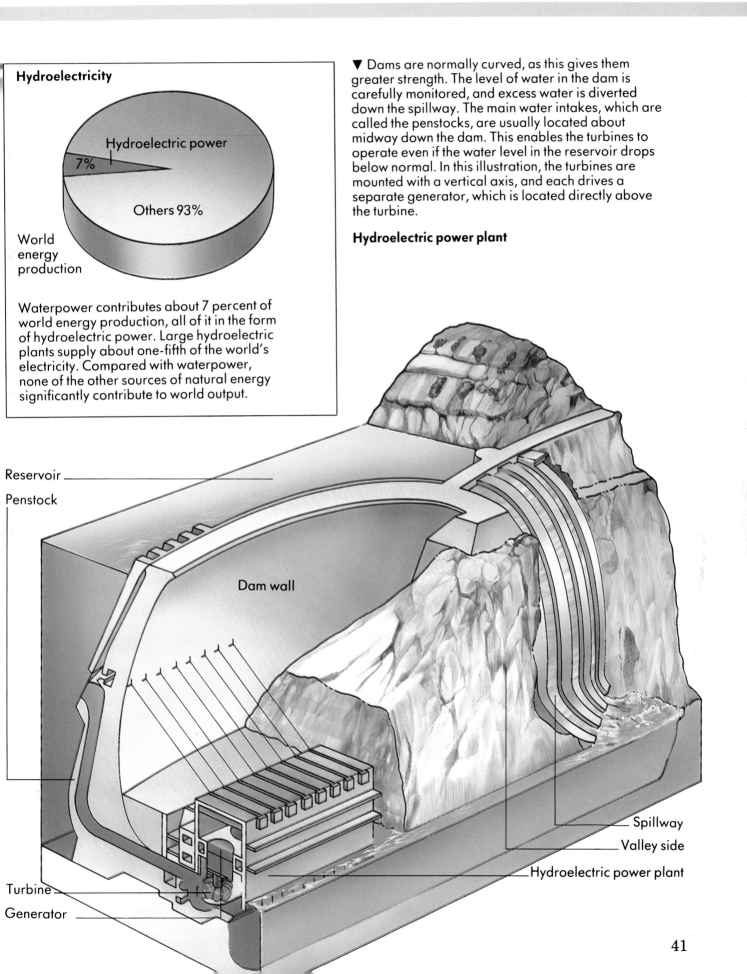

Hydroelectric power
7%

Others 93%

World
energy
production

Waterpower contributes about 7 percent of world energy production, all of it in the form of hydroelectric power. Large hydroelectric plants supply about one-fifth of the world's electricity. Compared with waterpower, none of the other sources of natural energy significantly contribute to world output.

▼ Dams are normally curved, as this gives them greater strength. The level of water in the dam is carefully monitored, and excess water is diverted down the spillway. The main water intakes, which are called the penstocks, are usually located about midway down the dam. This enables the turbines to operate even if the water level in the reservoir drops below normal. In this illustration, the turbines are mounted with a vertical axis, and each drives a separate generator, which is located directly above the turbine.

Hydroelectric power plant

Reservoir

Penstock

Dam wall

Spillway

Valley side

Hydroelectric power plant

Turbine

Generator

Wind power

Unlike water, the wind cannot be controlled or stored. Wind power must be exploited where and when it occurs naturally. Until very recently, the wind was used mainly to drive small pumps for agricultural purposes.

The amount of power produced by the wind increases as the cube of its velocity. This means that a doubling of the wind speed produces eight times as much power. In general, wind speeds increase with altitude. At 10 m (30 ft.) above the Earth's surface, the wind speed is about 20 percent greater than at ground level. At 60 m (200 ft.) up, it may be 50 percent greater.

Traditional windmills were designed to operate at fairly low wind speeds. The materials they were made from (wood and cloth) were not strong enough to withstand high winds.

Modern windmills, which are usually called wind turbines, are designed to operate at much higher velocities. As a result, they produce far more power, and can be used to generate electricity. There are two main types of wind turbine. The horizontal-axis turbine has the same basic layout as a traditional windmill. Instead of sails, it has a rotor shaped like an airplane propeller.

▼ Small wind turbines are used throughout the world to pump water and generate small amounts of electricity, particularly on farms. The commonest design uses a rotor consisting of a large number of metal vanes. The rotor turns on a horizontal axis. The whole of the turbine assembly is on a swivel mounting so that it can be turned to face the wind by the rudder.

► A wind farm in California, consisting of many rows of small Darreius wind turbines. Darreius turbines are easily recognizable by their distinctive shape, and can operate in wind coming from any direction. This particular wind farm is situated in a high mountain pass, where the winds are unusually strong and steady. Careful siting is the key to wind farming.

A vertical-axis wind turbine rotates on a shaft that is vertical to the ground. It normally has only two blades, mounted vertically at each end of a horizontal rotor.

The Darreius wind turbine is an advanced vertical-axis design that uses two curved blades. There is no separate rotor, and the blades are attached at each end of the shaft.

Wind turbines of both basic types are now in operation in many countries throughout the world. The largest ones are more than 100 m (over 300 ft.) tall. At full speed, the tips of the blades travel at up to 400 km/h (250 mph).

Wind turbines are positioned wherever the winds are strongest, and are often located on hills and cliff tops. As with solar power, there are two main approaches to the large-scale use of wind power. A wind farm is a large area of land containing many small wind turbines, up to 30 m (100 ft.) tall. Each of these contains a separate electrical generator. The other approach is to build just one or two very large turbines at each location. Taking maximum advantage of wind power may mean building wind farms offshore, where wind speeds are generally higher than over land.

Geothermal power

The Earth's crust is the thin, solid outer layer of our planet. Below about 100 km (60 mi.), heat from natural radioactivity is sufficient to keep rock in a softer state. Near the surface, the temperature rises generally about 30°C (over 50°F) for every 1,000 m (⅗ mi.) of depth. In areas of volcanic activity, this can increase to 80°C (nearly 150°F) per 1,000 m, and higher temperatures occur much closer to the surface.

In a few rare instances, in the United States, Japan, and Italy, this heat boils underground water, which rises to the surface as dry steam. This steam can be trapped and used to drive turbines. In California, the Geysers power plant has been built on top of a vast underground reservoir of dry steam. When operating at full capacity, the Geysers power plant supplies nearly as much electricity as is required by the city of San Francisco.

In most cases, superheated water remains trapped underground. When brought to the surface by wells, the water boils and the steam is used by turbines. Several countries, as far apart as Mexico and the Philippines, already produce electricity in this way.

Low-temperature geothermal heat has been used for thousands of years in public baths and health spas. In some parts of the world, hot water from volcanic springs and geysers is now a major source of domestic heating. In Iceland, more than two-thirds of the population now heat their homes with natural hot water. Other countries that make use of geothermal heating include the United States, the Soviet Union, China, Japan, France, and Hungary.

Even when there is no naturally-occurring underground water of suitable temperature, geothermal energy can still be harnessed. Hot, dry rocks may be used soon in many countries as a huge underground boiler.

▼ (right) The Geysers geothermal power plant in California supplies electricity to a city of half a million people. (left) This design is for a power plant producing electricity from hot dry rocks. Two wells are drilled some distance apart, one deeper than the other. The surrounding rock is then fractured with explosives to produce a large number of heat transfer surfaces. Cold water is pumped down the deeper well into the fissured rock, where it boils. Steam is tapped off by means of the other well.

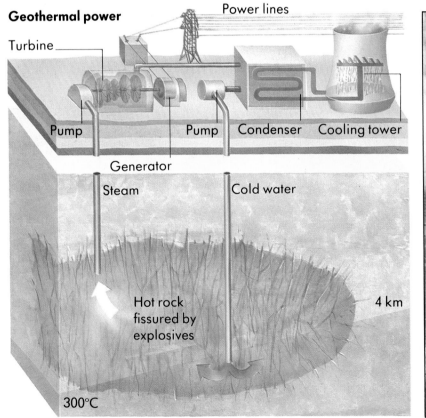

Geothermal power

Power lines

Turbine

Pump

Pump

Condenser

Cooling tower

Generator

Steam

Cold water

Hot rock fissured by explosives

300°C

4 km

Tidal and wave power

◀ A tidal-power dam across the estuary of the Rance River in France. At 750 m (820 yds.) long, the dam forms the world's largest tidal-power plant, with 24 turbines, which can operate when the tide is flowing in either direction.

▼ Wave energy converters: (1) The Salter "duck" uses a string of floating ducks hinged on a shaft; a full-sized duck would measure 25 m (82 ft.) across.(2) A development of the air-bag principle. The motion of the wave pumps air from one side of the converter to the other. The estimated length is 250 m (820 ft.). (3) Each wave-contouring raft would cover at least 5,000 square meters (6,000 sq. yd.).

1 Salter duck

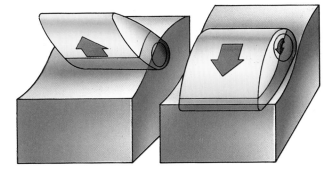

2 Air-bag device

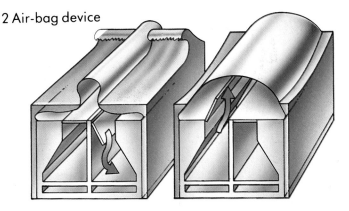

3 Wave-contouring raft

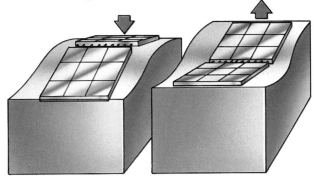

When the motion of the tides is channeled by a natural feature such as a river estuary, it produces a very strong flow of water. The power of the tides can then be harnessed by building a dam equipped with turbines across the tidal flow. Tidal-power dams have been built in France, Canada, the Soviet Union, and China.

The ceaseless up-and-down motion of ocean waves also represents a potentially valuable source of energy. A number of wave-power devices have been invented and tested as models, but none of them yet operate.

There are two basic approaches to wave power. The simplest designs place a line of wave-energy converters across the path of incoming waves. The energy converters float on the surface and consist of two sections hinged together. The motion of the waves operates the hinge, which powers pumps that drive turbines.

Other designs place the energy converter edge-on to the waves. The converters contain a number of air bags. As the wave travels along the converter, it squeezes the bags and forces air through turbines.

Part Two

Power Producers

Power and energy are not the same thing. Power is the ability to perform work. The energy we get from fuels and other sources must be transferred and transformed into useful power.

Electricity, produced by power stations and distributed by power lines, is the most useful commodity on Earth. Within the last hundred years, the supply of electricity has spread to nearly every part of the globe. In addition to providing basic heat and light, electricity powers hundreds of different home appliances and industrial machines.

Another vital form of power is provided by engines that drive machines. The first practical engines were steam engines, soon replaced in vehicles by the internal-combustion engines that power cars and trucks today. The invention of the jet engine enabled us to fly faster and further, and the rocket engine has enabled us to travel in space.

◀ Porcelain insulators on a pylon carrying high-tension electricity transmission lines in the Snowy Mountains, Australia. The electricity is produced by a hydroelectric power plant.

Power plants

Spot facts

• The largest and most powerful steam turbines can each produce enough electricity to supply nearly 800,000 houses.

• The condensers for the steam turbines of a 2,000-megawatt power plant require more than 200 million liters (over 50 million gallons) of cooling water every hour.

• Fuel-burning power plants have an average efficiency of about 35 percent: nearly two-thirds of the heat energy they produce is wasted.

• The hydroelectric power installation on the border between Brazil and Paraguay produces enough electricity to supply a city of as many as 35 million people.

► Tall and unsightly cooling towers are a characteristic feature of many power plants. They recycle cooling water for the condensers for the steam turbines. Hot water from the condensers is piped to the top of the towers and sprayed downward. By the time the water reaches the ground, it has cooled enough for it to be reused.

Electricity is produced by the constant high-speed rotation of turbine-driven generators. The story of electricity is also a story of turbines. The widespread use of electricity became possible only with the invention of the steam turbine in the late 1800s.

Today, power plants equipped with steam turbines generate about four-fifths of the world's electricity. The remainder is produced by water turbines in hydroelectric power plants. During recent years, there has been considerable interest in wind turbines as a source of electricity. This kind of power generation is less harmful to the environment. Many experimental designs have been built, but so far they make no significant contribution to world energy production.

The generation game

The useful properties of electricity were first demonstrated by the British physicist Michael Faraday. In 1821, Faraday built the first electric motor. Ten years later, he developed the principle behind the electrical generator, which is basically an electric motor in reverse. By the 1860s, small electrical generators powered by steam engines were being used in many parts of Europe and the United States.

The first generating stations were built to supply large individual buildings, such as hospitals and indoor markets. During the 1880s, demand for this new source of energy grew rapidly, and the first central generating stations were built. These were intended to supply electricity to the general public across a whole city district. In 1882, the American inventor Thomas A. Edison opened the famous Pearl Street power station in New York.

In order to produce electricity in commercial quantities, generators need to rotate at speeds of at least 1,000 rpm (revolutions per minute). Steam engines could just barely achieve that speed, and it meant operating to the limits of their capacity.

During the 1880s and 1890s, the British engineer Charles Parsons perfected the inward-flow steam turbine that could produce speeds of up to 18,000 rpm. Parsons's invention was rapidly adopted by the new electrical industry, and the first large steam turbines were installed in a German power station in 1901. Since then, steam turbines have become our most important source of electrical power.

Most power stations produce electricity in a two-stage process. Fuel of whatever kind, whether it be coal, oil, gas, or uranium, is first used to produce steam. The steam is then used to drive a turbine, which in turn produces useful quantities of electricity.

▼ The generator room at the Paris Opera House, installed by Thomas Edison in 1887. Steam engines worked flat out to provide high speeds of rotation. The rotary power was transferred to the generators by a series of belts. Today's power plants have turbines and generators mounted in line, and sharing a common rotating shaft.

▼ The first electrical generator built by Michael Faraday. Electrical current was produced by a copper disk spinning between the poles of an electromagnet. Modern generators use coils of copper wire instead. The first generators produced direct current. The switch to alternating current was made during the 1890s.

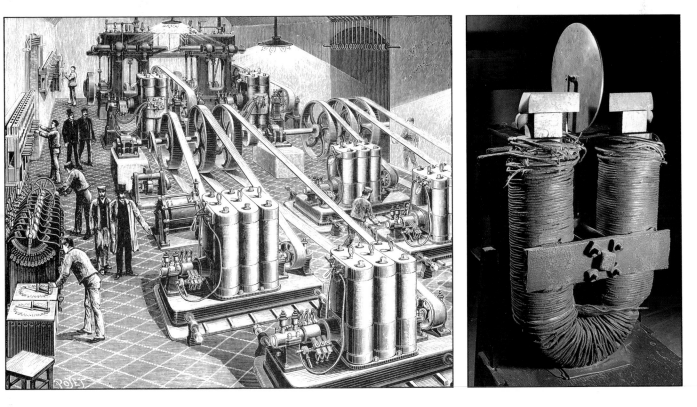

Steam turbines

A steam turbine consists of a central shaft, or rotor, mounted horizontally within a cylinder. The outer surface of the rotor is equipped with a large number of angled blades, which radiate like the spokes on a wheel. High-pressure steam is passed into the cylinder through a series of nozzles, mounted around its inner surface.

As the steam enters the cylinder, it expands. The heat energy of the steam is the kinetic energy (the energy of motion) of the vapor molecules. This energy is transmitted to the shaft through the angled blades, thus causing it to rotate. The speed of rotation depends on the temperature and pressure of the steam.

The introduction of new materials has enabled the construction of steam turbines that operate at extremely high temperatures and pressures. The design of the angled blades is also an important factor.

The simplest form of steam turbine is the impulse turbine, in which the blades are shaped like tiny cups. The steam releases all of its energy when it hits the turbine blades. The shaft rotates, but the steam itself comes to a dead stop.

A reaction turbine, of which Parsons's original turbine was an example, allows the steam to flow through a series of fixed and moving blades. These are designed to allow the steam to continue expanding as it passes through them. This means that the steam energy is used more efficiently, and the flow of steam is twice as fast as that of an impulse turbine. Modern steam turbines usually incorporate both types of blades in their design.

Still greater efficiency can be obtained by allowing the energy transfer to take place in a number of separate stages rather than all at once. This process, which is known as staging, is now used on all steam turbines.

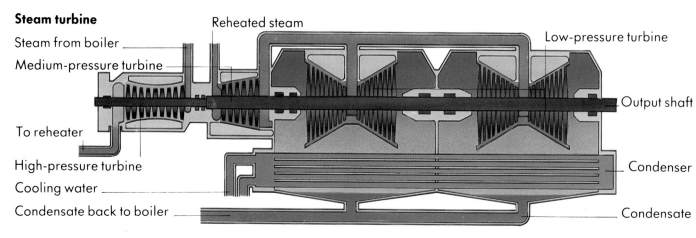

Steam turbine

Reheated steam

Steam from boiler

Medium-pressure turbine

Low-pressure turbine

Output shaft

To reheater

High-pressure turbine

Cooling water

Condensate back to boiler

Condenser

Condensate

▲ In a modern steam turbine, superheated high-pressure steam expands through progressively larger turbine blades. It is piped first into the high-pressure turbine, and, after reheating, into first the medium-pressure and then the low-pressure turbines. The steam condenses back into water in the condenser and is returned to the boiler.

▶ Steam turbines being assembled in a factory. In order to withstand constant exposure to high-temperature steam, the sets of blades have to made from extremely tough and resistant materials.

The commonest method of staging is to allow the steam to pass through a series of angled blades of progressively larger diameter. The steam drives the smallest set of blades first.

Large turbines take staging a step further by incorporating two or three separate sets of turbine blades mounted on the same shaft. Each stage is contained within a separate cylinder. The first stage usually has impulse blades and uses very-high-pressure steam.

After the steam has passed through the first stage, it is collected and reheated until it has sufficient energy to drive medium-pressure turbine blades. The steam is then piped into the low-pressure turbines. Turbines like these are known as compound turbines.

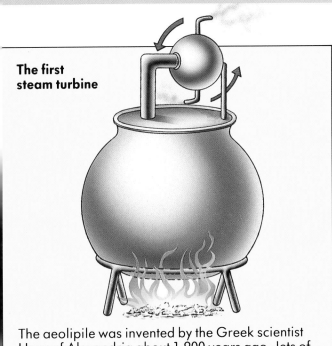

The first steam turbine

The aeolipile was invented by the Greek scientist Hero of Alexandria about 1,900 years ago. Jets of steam caused the ball to rotate.

▼ A large compound steam turbine of the type found in most power stations. The many sets of steam-driven blades are all attached to one central shaft.

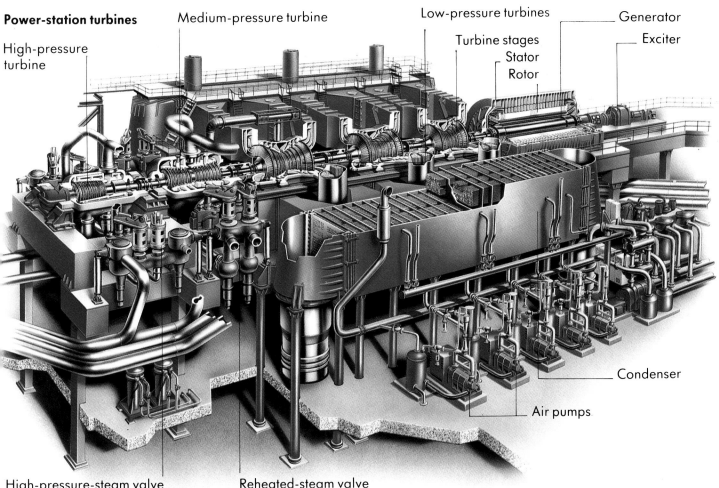

Power-station turbines

High-pressure turbine

Medium-pressure turbine

Low-pressure turbines

Turbine stages
Stator
Rotor

Generator

Exciter

Condenser

Air pumps

High-pressure-steam valve

Reheated-steam valve

51

Nuclear power

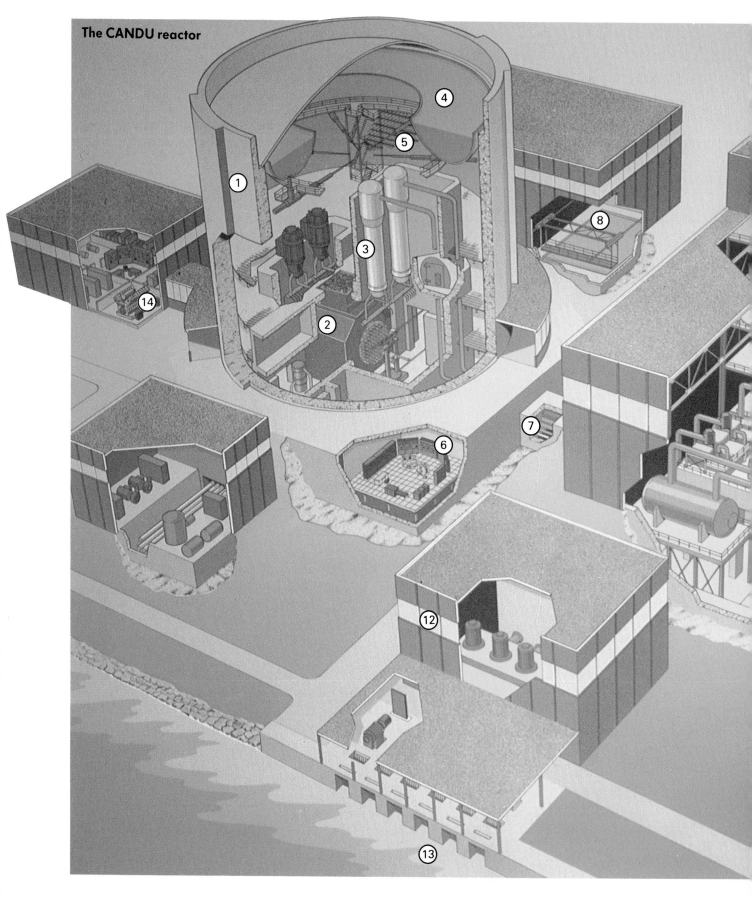

The CANDU reactor

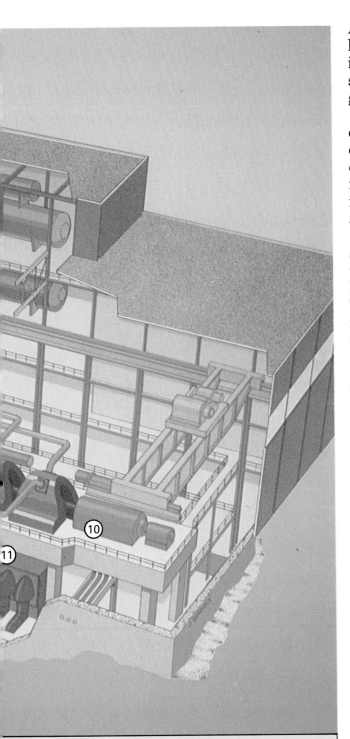

A nuclear power plant has the same basic layout as those fueled by coal, oil or gas. Water is heated in large boilers to produce steam. The steam is piped into turbines that turn generators, then it is cooled and condensed.

Although nuclear power plants do not produce any smoke or fumes, there is always a danger of radioactive gases leaking into the environment. The plants are usually sited in fairly remote areas. Within the power plant itself, every precaution is taken against the accidental escape of radioactivity.

The nuclear reactor itself is housed inside a separate building with very thick concrete walls. The reactor's coolant system transfers heat to heat exchangers that produce steam. The heat exchangers are also located in the reactor building. They are fed by an outside supply of water, and the steam leaves through underground pipes. This cuts to a minimum the time that the water and steam are exposed to radiation from the core.

Safety systems

The greatest danger is that the chain reaction in the core will get out of control, causing the reactor to overheat. A number of safety systems are operated to prevent this. When the temperature of the core starts to rise, control rods are automatically lowered into it to slow down the chain reaction. If necessary, they can be used to shut down the core completely. A secondary safety system, located above the reactor, can be used to drench the core with cold water. Apart from two major accidents, at Three Mile Island in the United States and Chernobyl in the Soviet Union, the nuclear power industry has a very good safety record.

◀ The CANDU (Canada Deuterium/Uranium) reactor was first used in 1971 at the Douglas Point power station in Ontario, Canada. It is widely considered to be one of the safest and most efficient, and has been exported to several countries including Argentina and India. The reactor building consists of a cylinder of prestressed concrete, lined with additional shielding and reinforcement. The emergency drenching system is located just below the domed roof. Cold water for the steam condensers in the turbine room may be taken from the nearby river or lake. Some nuclear power plants take cooling water from the sea.

Key	
Reactor building	7 Steam in underground pipes
2 Nuclear reactor	8 Storage for radioactive fuel
3 Heat exchangers	9 Turbines
4 Water storage for	10 Generator
emergency drenching	11 Water intake for condensers
5 Drenching sprays	12 Pump house for cooling water
6 Control room	13 Water outlet pipes
	14 Standby diesel generator

Wind and water turbines

Wind and water turbines work on the same principles as steam turbines, but operate at much lower pressures. The amount of energy in running water depends on the vertical distance that the water has fallen. This distance is known as the head. The greater the head, the greater the energy provided by a given quantity of water. In places where the head is above 30 m (100 ft), it is sufficient to produce a high-pressure water jet. Electricity can then be generated by using a simple impulse turbine, such as the Pelton wheel.

Most hydroelectric power stations use a type of turbine perfected by the American engineer James Francis in the mid-1800s. The Francis turbine is a reaction turbine. The angled blades are completely immersed in water, and are driven by the flow of the water through them, rather than by the impact of water against them. This enables the Francis turbine to operate with only a few meters of head, that is, with water with less pressure.

A Francis turbine consists of a single rotor, also called a runner, fitted with angled blades. The rotor is mounted within a spiral volute chamber. The turbine can be installed vertically or horizontally. In most cases, the rotor is mounted horizontally, and the electric generators are powered by a vertical shaft.

Where the head is too low for a reaction turbine, special low-head turbines can be used. One such design, the Kaplan turbine, uses a runner that is shaped like a ship's propeller. The Kaplan turbine can be installed within a volute chamber, or can be placed directly into the flow of water.

Wind turbines have been developed from the traditional windmill, and many different designs are now in use. Most installations use horizontal-axis turbines with rotors shaped like an airplane propeller. Vertical-axis wind turbines, such as the Darreius turbine, are less common. Their advantage is that they can operate whatever the wind direction.

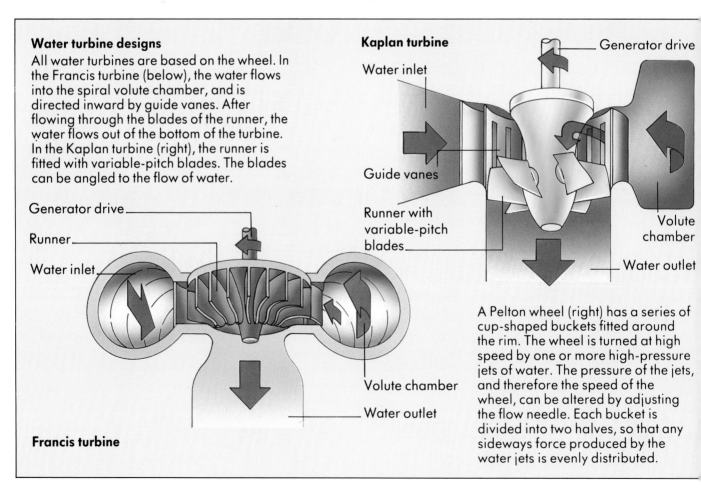

Water turbine designs
All water turbines are based on the wheel. In the Francis turbine (below), the water flows into the spiral volute chamber, and is directed inward by guide vanes. After flowing through the blades of the runner, the water flows out of the bottom of the turbine. In the Kaplan turbine (right), the runner is fitted with variable-pitch blades. The blades can be angled to the flow of water.

Generator drive
Runner
Water inlet
Volute chamber
Water outlet

Francis turbine

Kaplan turbine
Generator drive
Water inlet
Guide vanes
Runner with variable-pitch blades
Volute chamber
Water outlet

A Pelton wheel (right) has a series of cup-shaped buckets fitted around the rim. The wheel is turned at high speed by one or more high-pressure jets of water. The pressure of the jets, and therefore the speed of the wheel, can be altered by adjusting the flow needle. Each bucket is divided into two halves, so that any sideways force produced by the water jets is evenly distributed.

▼ The generator room at a hydroelectric power plant in the Snowy Mountains in Australia. The generators themselves are on the next level down, and the water turbines are on the next level below that. The power plant forms part of a mammoth hydroelectric and irrigation project. The project involved diverting the Snowy River back on itself so that it can water arid regions.

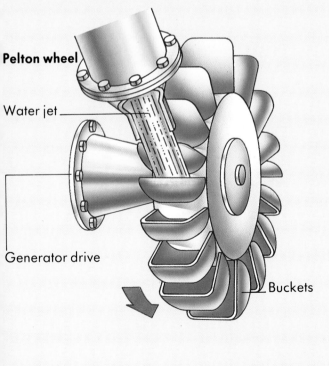

Pelton wheel

Water jet

Generator drive

Buckets

Rotor blades

Gearbox

Generator

Current flow

▲ A horizontal-axis wind turbine. Most designs use a three-bladed rotor, although some have only two. The main disadvantage of horizontal-axis turbines is that they can operate only when the wind is blowing straight at them.

Powerhouses

A power plant produces electric current in massive generators. These are often called turbogenerators, because they are driven by steam or water turbines. A more precise name is turboalternators, because they produce alternating current.

In power plants that use fossil or nuclear fuels, the turboalternators are located in a separate powerhouse. Steam for the turbines is sent along insulated pipes from the reactor or boiler room. Turboalternators operate at a constant speed, and this is adjusted by varying the pressure of the steam entering the turbines. In the United States, turboalternators typically operate at 3,600 revolutions per minute (rpm) to produce alternating current at 60 hertz (cycles per second). In Europe, they rotate at 3,000 rpm to produce current at 50 cycles.

The high speeds of operation mean that turboalternators also produce large amounts of unwanted heat. Modern designs are filled with hydrogen gas under pressure, which enables them to lose heat more efficiently. Water is circulated through the outer casing in order to carry the unwanted heat away.

Electricity is taken from each turbo-alternator by three cables in order to produce a three-phase supply. The cables run under the powerhouse floor to the bus room. In the bus room, electricity from the various turbo-alternators is brought together at a central point. The bus system usually consists of a network of heavy copper bars or cables, separated by large insulators.

Individual sections of the bus system are separated by circuit breakers. If there is a fault in any part of the system, these automatically stop the flow of current, and isolate the faulty section. The flow of electricity can also be regulated by heavy-duty switchgear, operated from the power plant's control room. Before electricity leaves the power plant. it passes through a series of transformers, which produce the very high voltages required for long-distance transmission.

In addition to the main turboalternators, most power stations also have smaller generators powered by gas turbines. These are used to produce additional electricity during periods when demand is especially high.

▲ The bus system and switchgear are protected against power surges by automatic circuit breakers.

◄ Supervisors in the control room constantly monitor the output of each turboalternator. They route the electricity through the bus system to the main transformers.

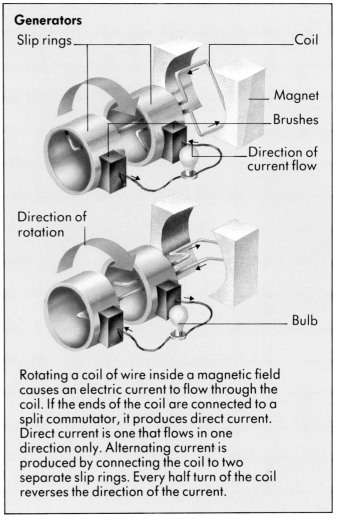

Generators

Slip rings — Coil

— Magnet

— Brushes

— Direction of current flow

Direction of rotation

— Bulb

▲ In the final stage of electricity production, the transformers step up the current to very high voltages.

▼ The powerhouse of a coal-fired power plant. Steam for the lines of turboalternators is piped up from the boiler room below. Cables carrying electric current to the bus room are set into the floor. Overhead cranes are used to lift up the massive outer casings of the turboalternators when they need servicing and maintenance.

Rotating a coil of wire inside a magnetic field causes an electric current to flow through the coil. If the ends of the coil are connected to a split commutator, it produces direct current. Direct current is one that flows in one direction only. Alternating current is produced by connecting the coil to two separate slip rings. Every half turn of the coil reverses the direction of the current.

Using electricity

Spot facts

• By 1900, more than two-and-a-half million light bulbs were being used to light London.

• In the United States, some power lines carry electricity at 765,000 volts.

• The power plants with the greatest output are hydroelectric. The U.S. plant with the most capacity is at Grand Coulee in Washington State, which has an output of 7.4 megawatts (million watts).

• The biggest ever power failure occurred in 1965 when an area of 200,000 square kilometers (80,000 sq. mi.) in eight northern U.S. states and Ontario was blacked out for up to 13 hours; 30 million people were affected.

Electricity is our most useful form of energy; it is instantly available at the touch of a switch. During the last 100 years, electricity has completely transformed human society. Electricity provides basic heating and lighting, and also powers a wide range of machines. In the developed countries, many homes now contain over a dozen different electrical appliances.

Power plants produce high-voltage electricity, which may be carried by power lines for hundreds of kilometers, to be used by factories and houses. Household appliances, however, use much lower voltages, and during the distribution of electricity, the voltage is progessively reduced. Inside the home, electricity is carried by a number of different circuits.

► Colored lights on the Christmas tree are now a familiar sight in many parts of the world. Traditionally, Christmas trees were lit with candles, which meant there was a constant risk of the tree catching fire. Electric lighting is much safer, but proper precautions must still be taken.

58

Transforming society

The single most important electrical device is the light bulb. Before its invention, the only sources of artificial lighting were oil and gas lamps. The first practical electric light bulbs were developed during the 1870s by Joseph Swan in Great Britain, and by Thomas Edison in the United States. Electric lighting had the advantage of being fairly cheap to install, and it was much safer than oil or gas. By 1900, electricity was in use throughout the world.

Other useful electrical appliances soon followed. By 1890, electrical engineers had perfected the electric space heater, with long-lasting heating elements made from nickel-chromium alloy. Electric stoves and vacuum cleaners first went on sale in the 1890s, but were not widely used before the 1920s.

Since 1920, electricity has completely transformed daily life, and a wide variety of household appliances are now available. Electric irons, food mixers, washing machines, refrigerators, air-conditioning units, and microwave ovens are now widely used.

Electricity also provided the energy for new forms of mass communication and popular entertainment. First radio and then cinema and television were made possible by the availability of electrical energy.

Industry too was transformed; electricity is now used in many basic processes, especially in the metalworking and chemical industries. More recently, electricity has allowed the development of electronic computers and the revolution in information technology.

◄ The Crystal Palace was built in London during the age of gas and oil lamps. In 1885 it was lit by electricity and provided a dramatic advertisement for the new energy source.

▼ The electric light bulb invented by Thomas Edison in 1879. The first light bulbs had a carbon filament made from cotton thread, which lasted for about only 40 hours. By 1900 there were longer-burning filaments made from metal wire. Modern bulbs use tungsten filaments.

Distribution

High-voltage power lines are used to carry large quantities of electricity over long distances. Energy is lost, mainly in the form of heat, as electricity travels down wires and cables. The use of very high voltages, typically ranging from 230,000 to 765,000 volts, reduces this energy loss to a minimum. The most efficient and widely used distribution method is by overhead power lines, supported by metal towers or poles. The atmosphere provides cooling, and most of the insulation. At points where the power line passes under the arm of a tower, it is encased in a glass or ceramic insulator.

Underground high-voltage lines are much more expensive to install, and are normally used only over short distances. The power line may be wrapped in paper impregnated with insulating oil and is contained in an oil-filled metal pipe. Gas-filled pipes are also used.

When electricity is taken from power lines, the voltage is stepped down by banks of transformers at a substation. Most substations also contain circuit-breakers and a variety of switchgear.

Stepping down the voltage to the levels used by the customer usually requires a series of substations. Factories often use electricity at 10,000 volts or more, which is much too high for domestic use. In cities, a series of local substations provide electricity for distribution to houses and stores. Large office buildings usually take electricity at a higher voltage than they can use directly, and have a separate substation on the premises.

In isolated regions, a power plant may be constructed to provide electricity directly to a town or large factory. In most cases, however, all the power plants in a country are linked into a national grid of power lines. Output from power plants can then be matched more closely to demand, and electricity can even be traded between different national grids. This normally takes place by overhead power line. Great Britain, however, imports some of its electricity from France along an undersea cable.

▼ Electricity is generated as alternating current at a power plant, where transformers step up its voltage to several hundred thousand volts. Power lines carried by tall towers transmit the high-voltage current to local substations, where transformers step down its voltage. Factories may take the current at over 10,000 volts. Homes use electricity usually at 120/240 volts.

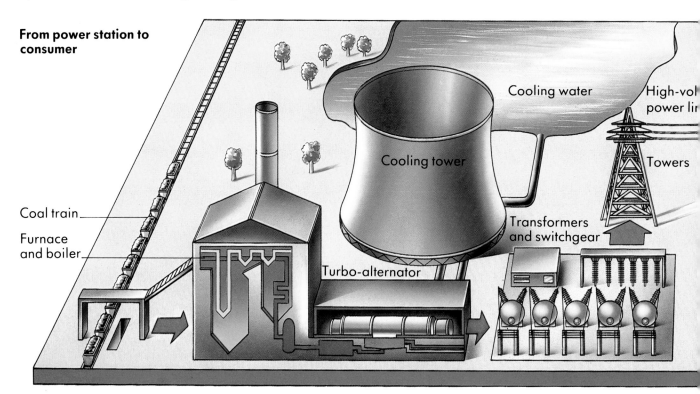

From power station to consumer

Coal train

Furnace and boiler

Cooling tower

Turbo-alternator

Cooling water

High-vol power lir

Towers

Transformers and switchgear

Transformers

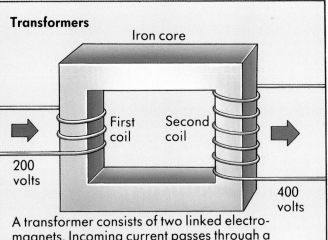

Iron core

First coil

Second coil

200 volts

400 volts

A transformer consists of two linked electro-magnets. Incoming current passes through a coil around the first electromagnet. The process of electromagnetic induction creates current in the coil around the second electromagnet. The voltage of the current is altered according to the number of turns in the two coils. If the second coil has twice as many turns as the first, then the voltage is doubled. If it has half as many, then the voltage is halved.

▲ A tower carrying high-voltage power lines, which here consist of sets of three cables. They have an aluminum conductor and are reinforced with steel.

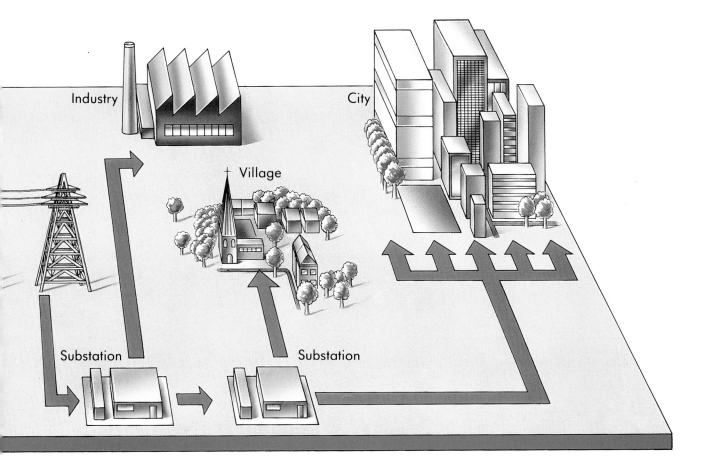

Industry

City

Village

Substation

Substation

Heating and lighting

Electricity gives off heat when it meets resistance to its flow. Useful quantities of heat can be produced by passing electricity through a coil of thin wire. A thin wire has a higher resistance to electricity than a thick wire. Winding the wire into a coil concentrates the heating effect into a small area. Coils used for heating are called heating elements, and are found in many household appliances. Maximum heat is produced when the element is glowing red hot.

An electric light bulb works by heating a coil of very thin wire, known as a filament, until it glows white hot. The first light bulbs contained a vacuum to prevent the filament from burning up. Modern bulbs are filled with an inert gas such as argon. A light bulb produces light across the whole range of the visible spectrum, but is very inefficient in its use of electricity. Only about 6 percent of the energy is released as useful light; the rest is lost as heat.

Discharge lamps are much more efficient at turning electricity into light. A discharge lamp consists of a glass tube containing a gas or vapor that conducts electricity. Metal contacts at each end of the tube allow current to pass through the gas. The flow of electricity causes the gas atoms to agitate, and as a result they give off light.

Different gases produce different colors of light. Sodium vapor produces orange light, and mercury vapor produces blue light. Ordinary discharge lamps are widely used for street-lighting, but the color of the light makes them unsuitable for use in the home or office.

Discharge lamps filled with neon gas can produce a wide variety of colors and are used in illuminated signs, often for advertising purposes. Other types of electric lamp are designed to produce infrared light, and are often used as bathroom heaters.

Fluorescent light tubes

Fluorescent tubes are special discharge lamps containing mercury vapor, and produce white light by means of a coating of phosphors on the inside of the tube. Besides visible blue light, mercury vapor gives off an invisible ultraviolet light. The phosphors absorb this and give off a white light which is suitable for most purposes.

Fluorescent tubes use much less electricity than ordinary light bulbs and last much longer. Small fluorescent tubes are now being incorporated in "economy" light bulbs that can be substituted for the ordinary type.

▶ The city of Boston by night. Electric lighting has greatly extended the working day, and offices and other buildings can now operate around the clock. Outside, powerful electric floodlighting also enables open-air activities, such as football games and other sporting events, to take place in the evening.

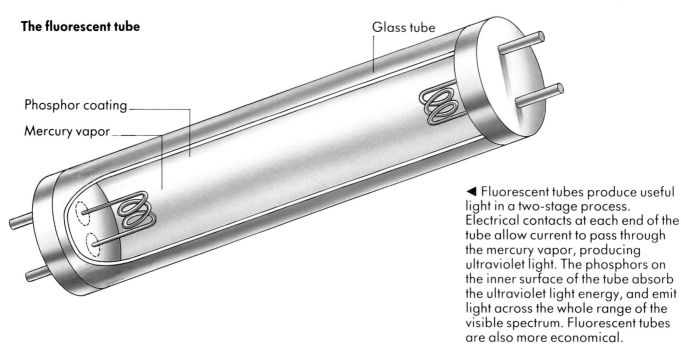

The fluorescent tube

Glass tube

Phosphor coating

Mercury vapor

◀ Fluorescent tubes produce useful light in a two-stage process. Electrical contacts at each end of the tube allow current to pass through the mercury vapor, producing ultraviolet light. The phosphors on the inner surface of the tube absorb the ultraviolet light energy, and emit light across the whole range of the visible spectrum. Fluorescent tubes are also more economical.

Heating by radio

A microwave oven heats food by means of high-frequency radio waves, known as microwaves, which have a very short wavelength. The microwaves are produced by a device known as a cavity magnetron. As the microwaves pass through an item of food, they cause its molecules to vibrate, and this vibration produces heat. The cavity magnetron was developed during World War 2 in order to produce tightly focused beams of microwaves for the first radar systems. Although the microwaves are said to form a continuous beam, the magnetron is in fact switching on and off about 10,000 times every second. The great advantage of microwave ovens is that they heat food extremely quickly.

Circuits and wiring

The electrical wiring inside a house consists of a number of branch circuits. The supply of electricity from outside goes to a watt-meter and then is split into the different circuits at the service panel, which has fuses to protect against the fire hazard caused by excessive current. The branch circuits use the voltage supplied by the service wires coming into the house, but are designed to carry different loads of electricity. Each circuit has a separate fuse, which melts if the circuit becomes overloaded. Some modern houses now use circuit breakers in the service panel instead of fuses.

A house has several general-purpose branch circuits – the number depends on the amount of floor space it has. There will also probably be at least a couple of circuits for heavy-duty appliances. In some cases a branch circuit is controlled by a two-way switch.

A circuit built into the ceiling supplies electricity to individual light fixtures. Light switches are usually wall-mounted and are connected to the circuit by longer loops of wiring.

Dangers of electricity

Electricity is dangerous. The current of electricity supplied to a house is high enough to kill people. Do not attempt to investigate or experiment with any of the electrical fittings in your home.

Domestic wiring

▼ Electricity used in the home comes from the local substation. The service panel, with fuses or circuit breakers is a safety device to isolate the different circuits from each other, and does not alter the voltage of the electricity in any way.

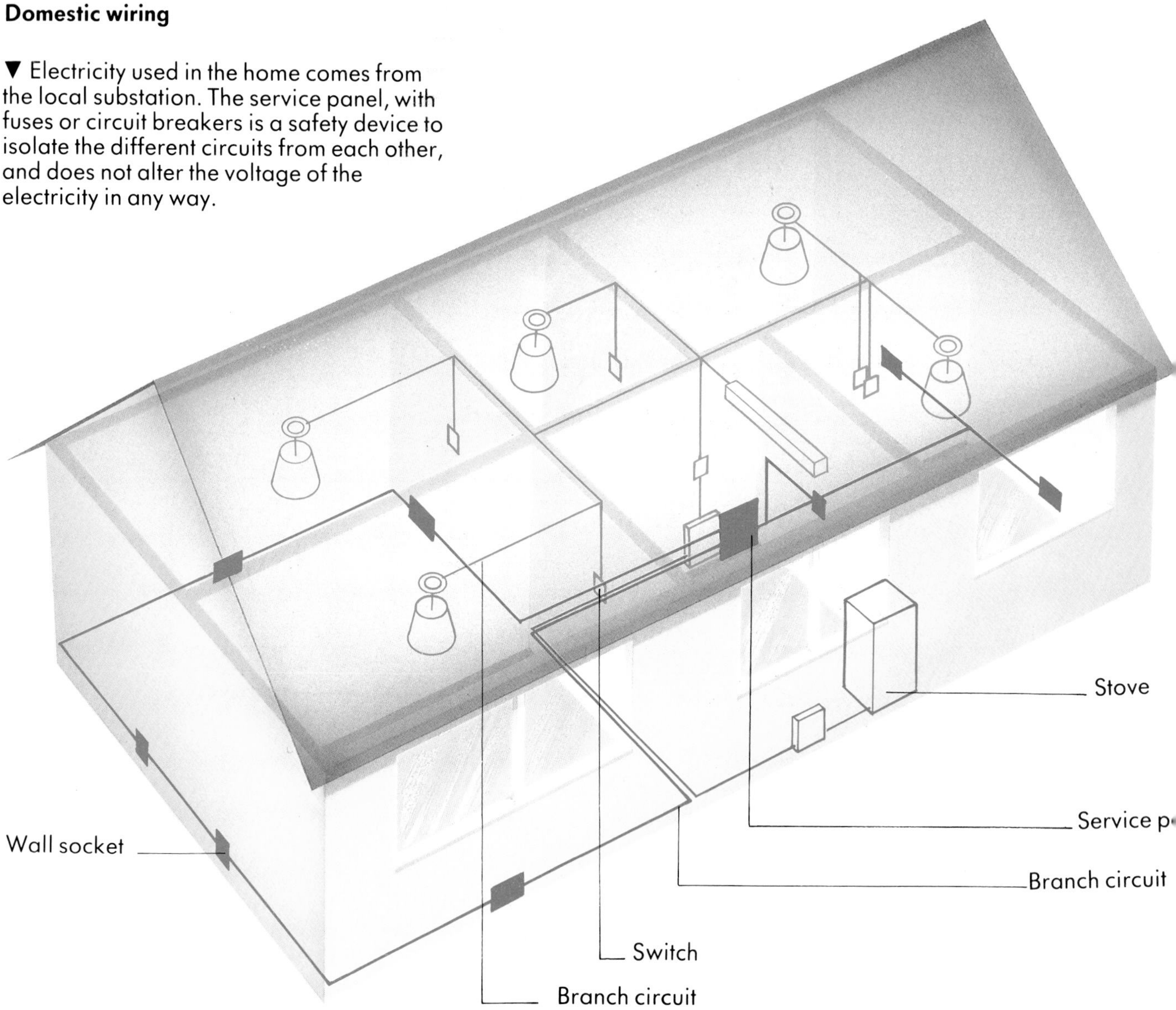

Stove

Service p

Branch circuit

Switch

Branch circuit

Wall socket

Motors

Electric motor

A direct-current motor consists of a coil of wire attached to a shaft and surrounded by simple magnet. When electricity flows through the coil, the effect of the magnetic field causes the coil to rotate. The supply of electric current to the coil is controlled by a split-ring commutator and brushes.

Apart from heating elements and lights, the most useful electrical devices are electric motors. Motors of many different sizes are built into many household appliances. Some use the motor's power directly, some indirectly.

Electric drills and food blenders make direct use of the motor's rotating shaft. Vacuum cleaners and hairdryers use motors to turn a fan which creates a flow of air. This can be used to pull air into the appliance, or blow air out.

Water pumps driven by electric motors are found in washing machines, shower units, and central heating systems. Refrigerators also use electric motors; in these appliances they circulate special coolant liquids.

▼ A hairdryer uses electricity in two different ways. A small electric motor drives a fan, and heat is produced by heating elements in the form of wire coils. As a safety precaution, the dryer is provided with a heat-sensitive bimetallic switch. If the dryer overheats, the switch opens, and breaks the electrical circuit.

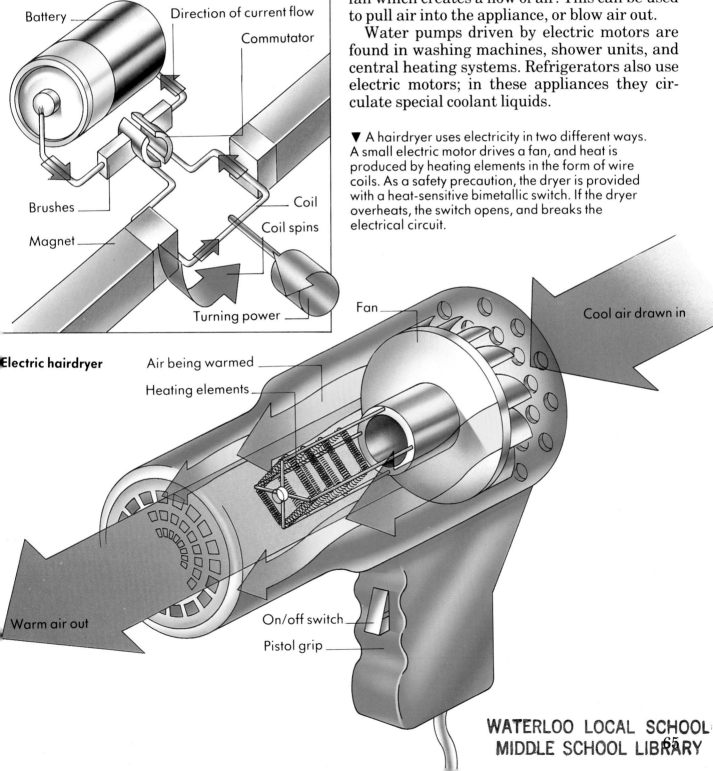

Battery — Direction of current flow
Commutator
Brushes
Magnet
Coil
Coil spins
Turning power

Electric hairdryer
Air being warmed
Heating elements
Fan
Cool air drawn in
Warm air out
On/off switch
Pistol grip

Engines

• A mass of steam has a volume at least 1,600 times greater than that of the same mass of water.

• China is the only country that is still building steam locomotives.

• One of the most powerful gasoline engines ever built had 28 cylinders arranged in four circles of seven, and was used in American bomber aircraft during World War 2.

• Diesel engines of over 3,000 horsepower are used to power many modern locomotives.

▶ A scale-model traction engine pulls a young visitor at a steam-power exhibition. Full-size traction engines were used in farming. In addition to providing motive power, they were used to drive agricultural machinery, by means of moving belts. Steam traction engines were also used as the first steamrollers, hence the name.

Engines are devices that burn fuel and provide mechanical power. Steam engines and gasoline engines both produce power through the up-and-down motion of a piston inside a cylinder. Engines that operate in this manner are called reciprocating engines.

Steam engines burn fuel separately to produce steam in a boiler. They are therefore known as external-combustion engines. Gasoline engines burn fuel inside the engine itself, and are internal-combustion engines.

The age of steam

f steam is condensed within a closed container, a vacuum is created. The first practical steam-powered device, invented by the Englishman Thomas Savery in 1698, was a vacuum pump.

In 1712, another Englishman, Thomas Newcomen, built the first true steam engine that used a piston within a vertical cylinder. Steam entering at the bottom of the cylinder forced the piston up. The steam was then condensed by a spray of cold water. The resulting vacuum allowed the piston to fall, ready to be driven up again by more steam. A connecting rod transferred the up-and-down motion to a hinged beam, hence the name "beam engine."

Newcomen's engines produced a steady pumping action, and were widely used in coal mines. But they were slow and inefficient.

Between 1769 and 1790, the Scottish inventor James Watt made a number of improvements, making the steam engine flexible and efficient. Watt's first improvement was the separate steam condenser, which allowed the cylinder to remain hot. He also invented double action, in which both the up-stroke and the down-stroke were steam-powered. Another improvement of his was to connect the piston to an offset flywheel, which converted the up-and-down motion into rotation.

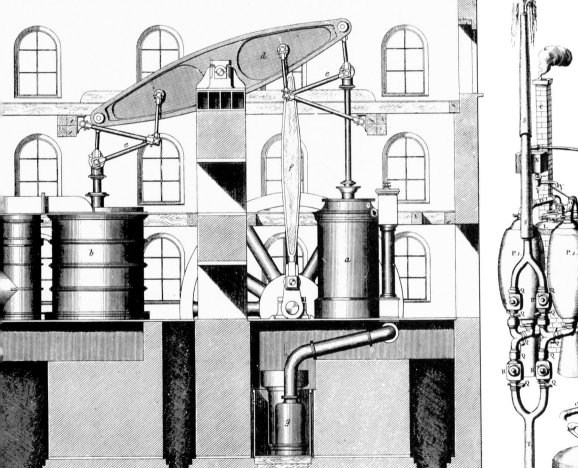

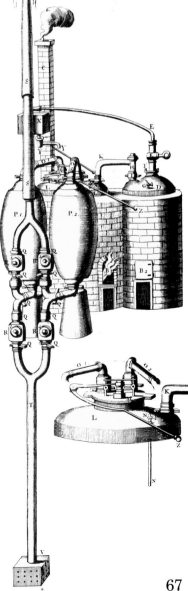

▲ A contemporary print showing an industrial steam engine of the mid-1800s. The right-hand arm of the beam (d) was pulled down when steam expanded against a piston in the steam cylinder (a) and was condensed in a condenser (g).

▶ Savery's steam pump used two boilers side by side and was very unsafe. Water was alternately boiled and condensed in each of the two boilers. The vacuum that was produced drew water up a pipe and through a one-way valve.

Steam engines

During the 1800s, the steam engine was further transformed. Vertical cylinders were replaced by horizontal ones, and better engineering techniques allowed high-pressure steam to be used with safety. This meant improved boilers, with steam being heated to progressively higher temperatures in a series of tubes within tubes. As a result, steam engines became smaller and produced more power.

Some designs recycled steam through a series of expansion chambers and condensers, thus making maximum use of steam's heat energy. Such engines tended to be both large and heavy, and were mainly used in factories.

Other designs abandoned bulky condensers and discharged steam straight into the atmosphere once it had been used. These smaller and lighter engines were used mainly in steam locomotives.

A steam locomotive is a self-contained unit that produces enough power to pull heavy loads along railroad lines at high speeds. Fuel, usually coal, is burned in the firebox to produce heat for the boiler. Steam locomotives normally have two cylinders, one on each side. The movement of the pistons in the cylinders is transferred to the driving wheels through a series of connecting rods and cranks.

Steam locomotive

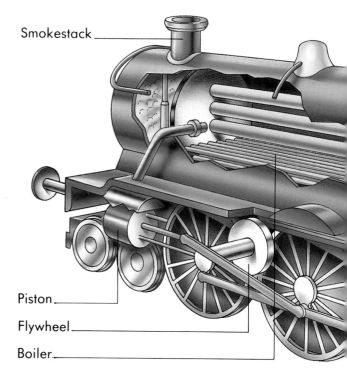

Smokestack

Piston

Flywheel

Boiler

generally obsolete with the development in the 1880s of the gasoline-engine automobile. Steam-powered traction engines, however, were used on farms and in road building for many years.

Stationary steam engines of all sizes were used throughout the 1800s to produce power in factories in Europe and North America. But by 1900 large steam engines were being replaced by the more efficient steam turbines, and smaller ones by internal combustion-engines, such as gas engines.

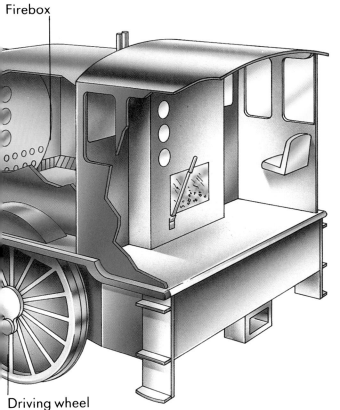

Firebox

Driving wheel

▲ In a locomotive hot gases from the firebox are drawn through the boiler tubes and boil the water. The steam is fed into the cylinders to drive the pistons, which are connected by rods to the drive wheels.

◄ A powerful steam traction engine, of the type used on farms from the late 1800s. They were used to pull plows and power machinery such as threshers.

The first steam locomotive, designed by the English engineer Richard Trevithick, appeared in 1804. By then steam power had also been applied to transportation on water. A paddle steamer called the *Charlotte Dundas* had been operating in Scotland since 1801. The first crossing of the Atlantic by a ship using steam power came in 1819. Later, steam carriages and cars enjoyed popularity, but they became

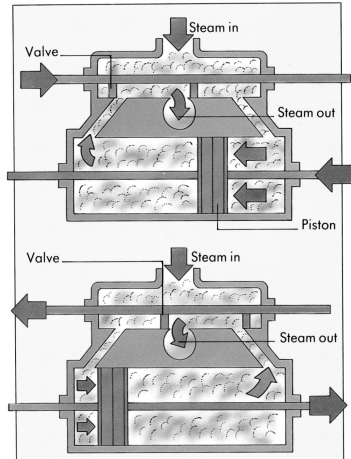

Double action

In a double-acting steam engine steam is introduced to each side of the piston alternately. In the diagram (top) the slide valve allows steam in on the right of the piston, and the steam forces it along the cylinder. The valve now moves across (bottom), allowing steam in on the left, and the spent steam escapes. James Watt introduced double action in 1781.

Gasoline engine

The gasoline engine is an internal-combustion engine. Fuel is burned inside a cylinder to provide the energy to drive a piston. Only one in four of the piston strokes produces power, and for this reason gasoline engines are known as as four-stroke engines.

The four-stroke engine was invented in 1876 by the German engineer N.A. Otto. The Otto engine burned a mixture of coal gas and air. In 1885, another German, Gottleib Daimler, invented the four-stroke gasoline engine. Many improvements were made by other engineers, and by 1900, the gasoline engine had acquired all the main features that it has today.

Gasoline engines burn a mixture of air and gasoline vapor in a ratio of about 14 parts air to one part gasoline. The mixture is compressed by a piston and is then ignited by an electric spark. The combustion, or explosion, pushes the piston down smoothly. The downward motion is transformed into rotation by the crankshaft.

Because of the four-stroke cycle, gasoline engines operate most efficiently when they have four cylinders operating in sequence. One of the cylinders is always delivering a power stroke, which means that the crankshaft is constantly receiving power. The rotating crankshaft is the source of all the useful power produced by the engine. Most of this power is used to perform work, for example propelling an automobile.

Gasoline engines are mainly used in automobiles and motorcycles. Many car engines have four cylinders, and larger engines often have multiples of four. The cylinders can be arranged in a line, or in a V-shape. Engines are often described as V-4, V-6, V-12, and so on. Motorcycles often have one or two cylinders.

Gasoline engines are capable of sustained operation at high speed. When producing maximum power, the pistons may travel up and down the cylinders 175 times every second.

◀ A high-performance racing car cornering at high speed. Many high-performance engines use superchargers or turbochargers to compress the air, so that a greater quantity of fuel can be burned at each stroke, thus producing more power.

Four strokes

The first stroke in the cycle (intake stroke) draws the gasoline/air mixture into the cylinder. The second stroke (compression stroke) compresses the mixture, which is ignited when the piston is at the very top of the up stroke. During the third stroke (power stroke) the piston is driven down by the rapidly expanding gases produced by combustion. The fourth stroke (exhaust stroke) forces exhaust fumes through the exhaust valve, and the cylinder is ready to repeat the cycle. Only the power stroke drives the piston. During the other three strokes, the pistons are driven by the rotation of the crankshaft.

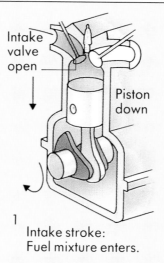

Intake valve open — Piston down

1 Intake stroke: Fuel mixture enters.

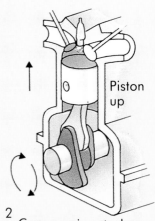

Piston up

2 Compression stroke: Mixture is compressed.

The gasoline engine

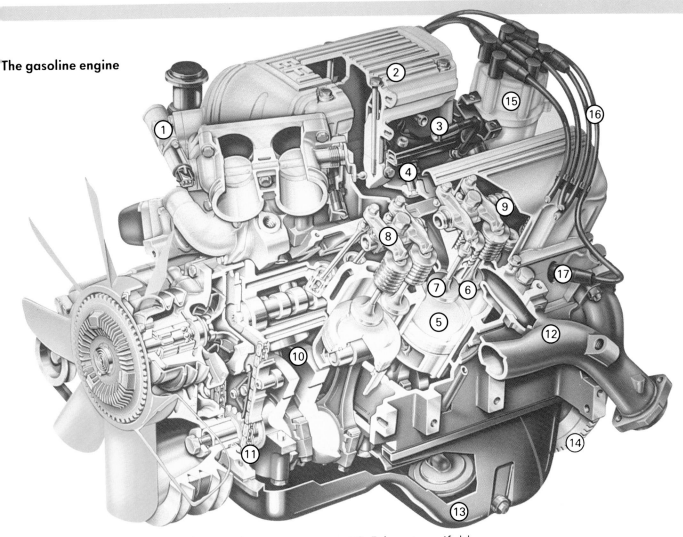

Key
1 Air intakes
2 Air chamber
3 Fuel line
4 Fuel injector
5 Piston
6 Exhaust valve
7 Intake valve
8 Valve rocker arm
9 Camshaft
10 Crankshaft
11 Chain drive for camshaft
12 Exhaust manifold
13 Oil pan
14 Flywheel
15 Distributor
16 High-voltage lead
17 Spark plug connector

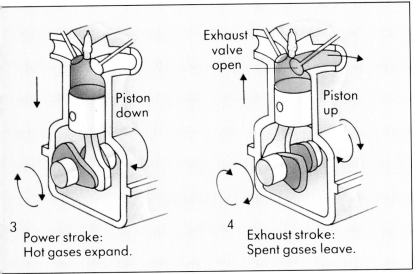

3 Power stroke:
Hot gases expand.

4 Exhaust stroke:
Spent gases leave.

▲ Cutaway drawing of a modern V-6 gasoline engine, typical of those used in medium-sized cars. The cylinders are arranged in two rows of three, set at about 120° to each other. Air enters the intakes and passes through the air chamber to the intake valves. Gasoline is pumped along the fuel line and is injected into the air flow and vaporizes. After combustion, exhaust fumes pass through the exhaust valves and into the exhaust manifold. Most of the power produced by the engine is transmitted by the crankshaft to the flywheel at the back of the engine. At the front, the crankshaft also drives both the cooling fan and the overhead camshaft. The camshaft operates the valves by means of a series of rods connected to spring-loaded rocker arms.

Engine systems

In addition to fuel and air, a gasoline engine requires electricity, oil, and water. These are supplied by different systems.

The fuel system pumps liquid gasoline from the fuel tank and distributes it in measured quantities into the air flow to the cylinders. Many engines regulate the flow of gasoline with mechanical devices known as carburetors. Some modern engines now use electronically-controlled fuel injectors. The engine's breathing system draws in air through a filter, and channels the air flow to the intake valves.

The exhaust system takes hot gases from the cylinders through the exhaust manifold and out of the engine. In most cases, a muffler is located at the end of the exhaust system. Some countries also require that exhaust gases pass through a catalytic converter to reduce the amount of pollution emitted.

The engine's lubrication system ensures that all moving parts are coated with a thin film of oil. This reduces friction. The main oil reservoir is the oil pan, which is located at the base of the engine beneath the crankshaft.

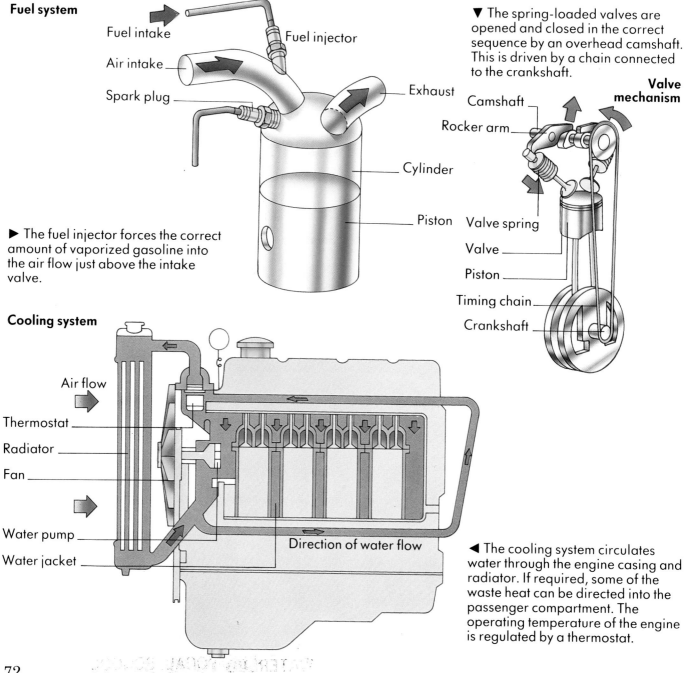

Fuel system

Fuel intake

Air intake

Spark plug

Fuel injector

Exhaust

Cylinder

Piston

▶ The fuel injector forces the correct amount of vaporized gasoline into the air flow just above the intake valve.

▼ The spring-loaded valves are opened and closed in the correct sequence by an overhead camshaft. This is driven by a chain connected to the crankshaft.

Valve mechanism

Camshaft

Rocker arm

Valve spring

Valve

Piston

Timing chain

Crankshaft

Cooling system

Air flow

Thermostat

Radiator

Fan

Water pump

Water jacket

Direction of water flow

◀ The cooling system circulates water through the engine casing and radiator. If required, some of the waste heat can be directed into the passenger compartment. The operating temperature of the engine is regulated by a thermostat.

► All the moving parts in an engine receive constant lubrication. A pump draws oil from the oil pan and forces it under pressure through channels to the engine bearings on the crankshaft and to the camshaft. Other areas get splashed with oil. Despite lubrication, some engine wear takes place. For this reason the lubrication system incorporates a filter to remove solid fragments before they do damage.

▼ A spark plug creates an electric spark across a tiny gap between two metal contacts, or electrodes. The inner electrode is connected with the plug lead and receives high-voltage current from the ignition coil via the distributor. The outer, or ground, electrode is connected to the plug casing. The two electrodes are insulated from each other by a thick ceramic layer.

Spark plug

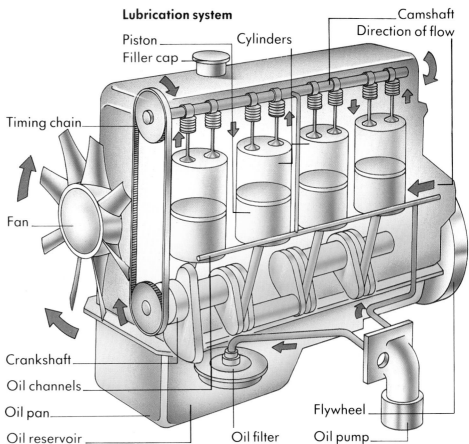

Lubrication system

Piston · Filler cap · Cylinders · Camshaft · Direction of flow · Timing chain · Fan · Crankshaft · Oil channels · Oil pan · Oil reservoir · Oil filter · Flywheel · Oil pump

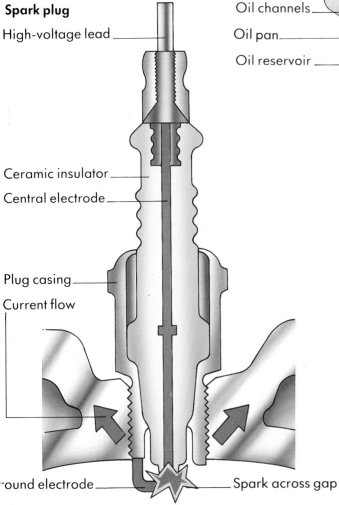

High-voltage lead
Ceramic insulator
Central electrode
Plug casing
Current flow
Ground electrode — Spark across gap

Some engines are designed to be air-cooled. They radiate away unwanted heat directly into the atmosphere. Air-cooled engines, such as many motorcycle engines, usually have a series of cooling fins around the cylinder casing.

Most gasoline engines use both air and water to carry heat away from the cylinders. Water circulates through the engine and radiator inside a sealed system. Heat from the engine heats the water, which is then cooled inside the radiator. A stream of air is drawn through the radiator by a fan mounted at the front of the engine. The fan is usually turned by a belt attached to the crankshaft.

The electrical system is based around a low-voltage (12-volt) battery. The battery supplies current to the ignition coil, which produces the high voltage needed by the spark plugs. High-voltage current is supplied to each spark plug in turn by the distributor, which is a kind of high-speed rotary switch. The battery is kept charged by a generator driven by the crank-shaft. The electrical system may also be used to drive the fuel, oil, and water pumps.

Other engines

Other types of piston engine share the same basic design as the gasoline engine, but differ slightly in the details of their operation. Some, such as the diesel engine, are designed to use a different type of oil-based fuel.

The diesel engine was invented by the German engineer Rudolf Diesel in 1893. Like the gasoline engine, the diesel engine is a four-stroke engine. It differs, however, because it has no separate ignition system. The diesel engine is self-igniting, and uses the principle that air heats up when it is compressed. The action of the piston in a diesel engine compresses the air so much that it becomes hot enough to ignite the fuel.

Diesel engines are more efficient than gasoline engines, but tend to be quite heavy. They are mainly used in large vehicles, such as trucks, buses, agricultural vehicles, and railroad locomotives. Small diesel engines are now found in some cars. Diesel engines are also widely used to power electrical generators.

▲ Trucks powered by large diesel engines are used in Australia to pull road trains, consisting of several trailers. Once the road train gets up to its cruising speed, the diesel engine performs very efficiently. At slow speeds, however, this form of transportation is extremely expensive because of its high level of fuel consumption.

◄ Motorcycle riding is popular in many countries, especially with the young. Two-stroke engines have a much simpler design and fewer moving parts than four-stroke engines. Though not as powerful, they are generally more reliable. This makes them ideal for young riders who like to do their own servicing and maintenance.

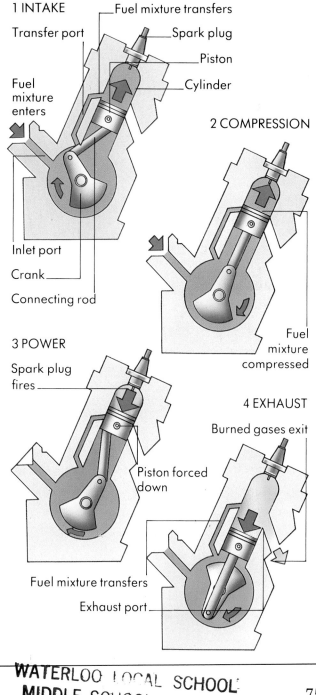

Intake and exhaust are controlled by the piston. The fuel mixture of gasoline, oil, and air enters the inlet port and then transfers to the cylinder. The upstroke of the piston closes the transfer port; the fuel mixture is compressed and then ignited by a spark plug. During the downstroke (the power stroke), exhaust gases leave through the exhaust port, and fresh mixture flows in through the transfer port.

1 INTAKE Fuel mixture transfers

Transfer port Spark plug

Piston

Fuel mixture enters Cylinder

2 COMPRESSION

Inlet port

Crank

Connecting rod

Fuel mixture compressed

3 POWER

Spark plug fires

4 EXHAUST

Burned gases exit

Piston forced down

Fuel mixture transfers

Exhaust port

The two-stroke engine is a gasoline engine that produces power every other stroke. Two-stroke engines have no separate lubrication system, and lubricating oil has to be mixed into the gasoline. Combustion is less efficient, and large amounts of exhaust fumes are produced.

Some gasoline engine designs do not use a piston within a cylinder. Rotary engines, such as the Wankel engine, use a flat, three-cornered rotor that turns around a central shaft. It revolves in a chamber shaped like a figure eight. As it revolves, it creates spaces in which the four stages in the four-stroke engine cycle take place. These four stages are intake, compression, power, and exhaust.

The basic gasoline engine can also be adapted to run on other fuels. During World War 2 many cars and trucks ran on gas. More recently, Brazil has tried out alcohol-powered cars.

Jets and rockets

▶ A Harrier "jump jet" uses swiveling nozzles to direct the exhaust gases from its jet engine downward as it takes off and lands vertically. It swivels the nozzles through 90° to produce horizontal thrust so that it can fly normally.

Jets and rockets produce power in the form of direct thrust. They burn fuel and concentrate a stream of exhaust gas through a nozzle at the back of the engine. This backward flow of hot gas causes the engine to move forward by reaction, and for this reason jets and rockets are called reaction engines.

Jet engines breathe air, and are now used in all types of aircraft and helicopters. Rockets carry their own oxygen supply and can produce far more power than jet engines, but can burn fuel for only a fairly short period of time. During the period of combustion, rockets can achieve very high velocities: fast enough to escape the pull of Earth's gravity and travel into space.

The jet age

Until the 1950s, most aircraft were propeller-driven. The rotary motion needed to turn the propellers was provided by high-performance gasoline engines. The most powerful engines could propel a single-seater aircraft up to about 660 km/h (410 mph). Higher speeds required a different kind of engine, and by the 1930s many aircraft engineers were thinking about jet propulsion.

The invention of the jet engine is generally credited to Frank Whittle, an officer in the British Royal Air Force (RAF). Whittle patented the basic design of the jet engine in 1930, but independent research was also taking place in other countries. The first jet-powered flight was made by an experimental German aircraft, the Heinkel He-178, in August 1939.

Whittle's design was for what we today call a turbojet. Kerosene fuel is burned inside a stream of compressed air. The expanding exhaust gases drive a turbine before leaving the engine and providing thrust. Such jet engines are thus gas-turbine engines.

Toward the end of World War 2, Whittle's engine was used in the Gloster Meteor, the first jet aircraft to enter regular service. The Meteor, which became operational in 1944, was a twin-engined jet fighter, with one engine in each wing. The first jet-propelled passenger aircraft was the De Havilland Comet, which made its maiden flight in 1949, and entered service in 1952. Other passenger jets, including the Boeing 707, soon followed. By the 1960s, jets had largely replaced gasoline-engined aircraft.

The main advantage of the jet engine is that it produces far more power than a gasoline engine of the same size and weight. Jet engines can therefore propel larger aircraft at higher speeds over greater distances. Another advantage is they are mechanically simpler, producing rotary motion directly without the need for connecting rods and cranks. They also run more smoothly, with less vibration. And whereas the efficiency of propellers decreases with altitude, that of jets increases. Many aircraft now use turboprop and turbofan engines.

▼ Frank Whittle in 1944, the year Great Britain's Gloster Meteor jet fighter went into service with the RAF. The first flight of a plane with a Whittle jet engine was in 1941, two years after the first jet plane, the He-178.

▲ A modern turbofan engine, of the type used to power most of today's airliners. It differs from the turbojet of the Whittle design by having a huge fan in front. The fan not only directs air through the engine, but also directs it around the engine. For this reason it is sometimes called a by-pass turbojet.

Jet engines

The simplest form of jet engine is the ramjet, which has no compressor or turbine. A ramjet relies on the speed of its own motion through the atmosphere to compress the air entering at the front of the engine. Fuel is burned at the center of the engine, and the expanding gases are directed through a nozzle at the rear.

The main disadvantage of the ramjet is that it cannot operate when the engine is not moving. Ramjets are therefore used only in missiles that are launched by a rocket motor. Once the missile is moving, the ramjet takes over.

A turbojet uses a rotary compressor at the front of the engine to provide high-pressure air to the combustion chamber. The exhaust gases pass through a turbine before leaving through the exhaust nozzle. The turbine provides just enough power to drive the compressor by means of a central shaft.

The power output of a turbojet can be boosted for short periods by afterburning. Additional fuel is burned in a second combustion chamber, which is located between the turbine and the exhaust nozzle.

▲ Turbojets are now mainly used in high-performance aircraft such as this two-seater Alphajet. The air intakes are situated on either side of the fuselage just below the cockpit canopy.

◀ The turbojet consists of a central combustion chamber with a compressor in front and a turbine behind. Modern engines use a multistage compressor to force air into the combustion chamber.

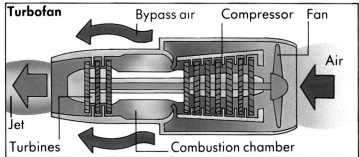

Turbofan

Bypass air · Compressor · Fan · Air · Jet · Turbines · Combustion chamber

▲ The turbofan engine has a propeller-like fan in front, which forces air around, as well as through, the compressor and combustion chamber. The fan is driven by a turbine in the jet exhaust. The bypass air joins the gases in the exhaust to produce thrust.

◄ With wing flaps and landing gear down, an Airbus A310 airliner of Switzerland's national airline, Swissair, comes in to land. The plane has two turbofan engines.

► Four turboprop engines power this curious-looking aircraft, known as the Super Guppy. They give it a maximum speed of about 460 km/h (nearly 290 mph).

▼ In the turboprop engine the propeller is mounted outside. It is driven, through reduction gearing, by a separate set of turbines in the jet exhaust.

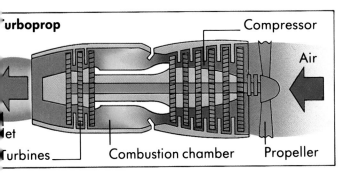

Turboprop

Compressor · Air · Jet · Turbines · Combustion chamber · Propeller

A turboprop engine operates in much the same way as a turbojet, but has two sets of turbines. One set drives the compressor, while the other drives a propeller mounted at the front of the engine. Although most propulsive thrust comes from the propeller, some comes from the gases exhausting from the turbines.

Turboprops are more efficient than turbojets at low speeds, and are widely used in multi-engined transport aircraft. The rotary motion this type of engine produces also makes it suitable for helicopters. In the helicopter, it is known as a turboshaft engine.

The most common engine used in passenger-carrying airliners is the turbofan, which is a development of the turboprop. In a turbofan engine, the propeller, or fan, is mounted inside the engine casing in front of the compressor. Part of the air drawn in by the fan is directed into the compressor and combustion chamber. But most is directed around them, straight into the jet exhaust. About 75 percent of the engine thrust is produced by the fan. Turbofans are also called ducted-fan or bypass turbojet engines. They are quieter than turbojets and turboprops and consume less fuel.

Rocketry

Rockets were invented in China more than 800 years ago. The first rockets were fueled by gunpowder, just like modern fireworks, and were used as weapons. The rocket was revived as a weapon of war in the late 1700s. Rockets were used by the British against the Americans in the War of 1812. The event is commemorated by the words "rockets' red glare," in the American national anthem.

By the end of the 1800s, scientists in many countries were considering rockets as a means of propulsion. In 1895 the Russian scientist, K. E. Tsiolkovski, was the first to propose rockets as a means of space travel. He also foresaw the use of liquid rocket fuels, and invented the important concept of the multistage rocket. For many years, however, Tsiolkovski's work received little recognition.

In the early 1900s, rocket research was taken up by the American scientist and inventor, Robert H. Goddard. In 1926 Goddard successfully launched the world's first liquid-fueled rocket. Although it traveled less than 60 m (200 ft.), it was an important first step.

During World War 2, German scientists turned the liquid-fueled rocket into a long-range weapon. The V-2 rocket was first launched in 1944, and could carry a 1-ton explosive warhead for more than 300 km (nearly 200 mi.). Later, much bigger rockets were developed both by the United States and Soviet Union.

By the early 1960s, both countries had developed large, multistage rockets. These were powerful enough to escape Earth's gravity, and could be used to lift astronauts and artificial satellites into space.

▲ One of the huge main engines fitted to the first stage of the Saturn V rocket. The large bell-shaped structure is the combustion chamber. Liquid fuel and liquid oxygen are supplied by pipes, valves, and pumps mounted above the combustion chamber.

▶ The American scientist Robert H. Goddard standing alongside the first liquid-fueled rocket in March 1926.

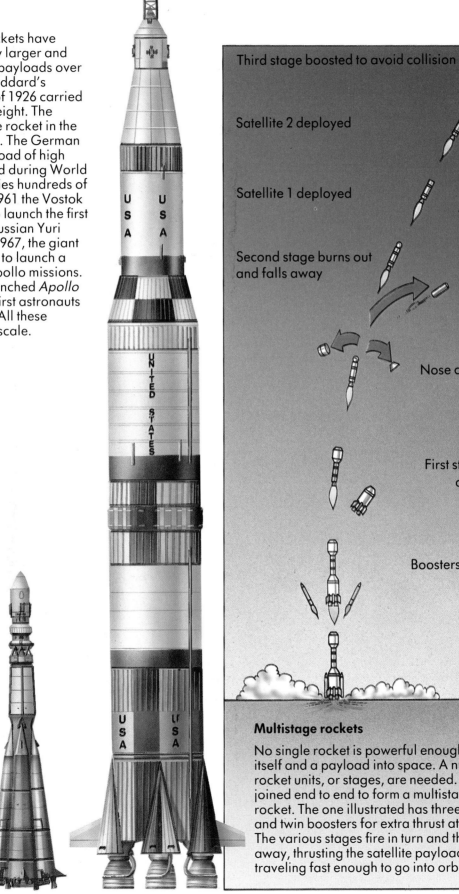

▶ Over the years, rockets have become progressively larger and have carried heavier payloads over greater distances. Goddard's experimental rocket of 1926 carried nothing but its own weight. The diagram (1) shows the rocket in the photo, drawn to scale. The German V-2 (2) carried a payload of high explosives. It was used during World War 2 to bombard cities hundreds of kilometers away. In 1961 the Vostok rocket (3) was used to launch the first man into space, the Russian Yuri Gagarin. Starting in 1967, the giant Saturn V (4) was used to launch a series of American Apollo missions. In 1969 a Saturn V launched *Apollo 11* which carried the first astronauts to land on the Moon. All these rockets are drawn to scale.

Third stage boosted to avoid collision

Satellite 2 deployed

Satellite 1 deployed

Second stage burns out and falls away

Nose cone jettisoned

First stage burns out and falls away

Boosters separate and fall away

Lift-off

Multistage rockets

No single rocket is powerful enough to lift itself and a payload into space. A number of rocket units, or stages, are needed. They are joined end to end to form a multistage rocket. The one illustrated has three stages and twin boosters for extra thrust at lift-off. The various stages fire in turn and then fall away, thrusting the satellite payload until it is traveling fast enough to go into orbit.

1   2   3   4

81

Liquids and solids

Steam engines, gasoline engines, and jets are all air-breathing engines. They take the oxygen required for combustion from Earth's atmosphere, and need only to be supplied with fuel. Rocket engines, however, carry their own supply of oxygen as well as their fuel, which gives them two important advantages. The performance of a rocket engine is not limited by the amount of air that it can take in. This means that rockets can produce far more thrust than jet engines of the same size. More importantly, carrying their own oxygen means that rockets can operate outside the atmosphere, in the near-vacuum of space.

There are two main types of rocket: those that use solid fuel, and those that use liquid fuel. Solid-fueled rockets, of which gunpowder rockets are the commonest example, carry fuel and oxygen mixed in solid chemical form. Once the fuel has been ignited, it burns steadily until it has all been consumed. The advantage of solid-fueled rockets is that once manufactured, they are ready for immediate use. They are mainly used in missiles, and as launch boosters for bigger rockets.

Liquid fuels

Liquid-fueled rockets carry both fuel, such as hydrogen, and oxygen in liquid form, inside separate refrigerated tanks. Fuel and oxygen are injected into the combustion chamber at high pressure and are then ignited.

The main advantage of liquid-fueled rockets is that they can be shut down during flight, and then reignited when required. The power output of the engine can also be adjusted during flight by regulating the amount of fuel and oxygen passing through the pumps and injector. The main disadvantage of such rockets is that fuel and oxygen cannot be stored on board. The fueling operation, which can take several hours, must be carried out shortly before the rocket is to be launched. Liquid-fueled rockets are used in the largest intercontinental missiles and for space exploration.

▼ The complex flow of fuel and oxygen through a typical liquid-fueled rocket. Temperatures inside the combustion chamber can reach over 5,000°C (9,000°F), and some of the liquid hydrogen is used to cool the combustion chamber.

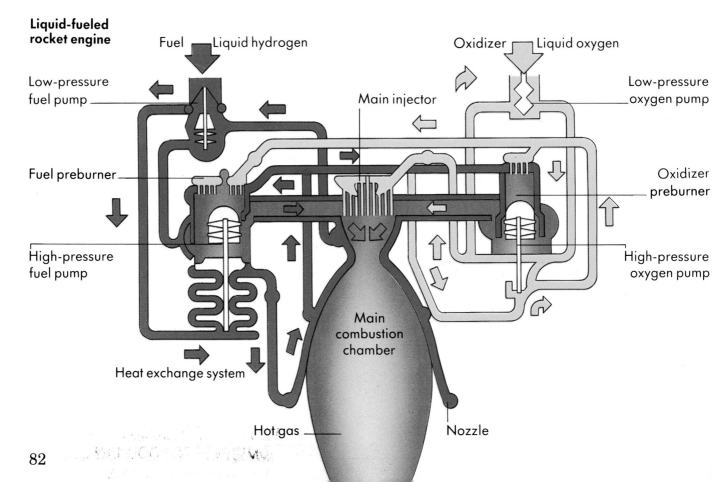

Liquid-fueled rocket engine

Fuel — Liquid hydrogen

Oxidizer — Liquid oxygen

Low-pressure fuel pump

Main injector

Low-pressure oxygen pump

Fuel preburner

Oxidizer preburner

High-pressure fuel pump

High-pressure oxygen pump

Main combustion chamber

Heat exchange system

Hot gas

Nozzle

▲ Building one of the space shuttle's booster rockets inside the Vehicle Assembly Building. The exhaust nozzles are visible at the bottom. Most of the booster's mass consists of a tall column of solid fuel.

▲ Inside the Vehicle Assembly Building at the Kennedy Space Center in Florida, the space shuttle orbiter *Discovery* is lowered into position on its external fuel tank and solid rocket boosters. The external tank provides propellants for the orbiter's three main liquid fueled engines.

▶ A battery of Nike-Hercules missiles. The Nike-Hercules missile has a solid-fueled rocket engine and is designed to shoot down aircraft at altitudes up to 30 km (nearly 20 mi.) above the Earth's surface. Liquid-fueled rockets could not be used for this purpose. They take too long to fuel and prepare for launch to be used in an emergency.

Rocket workhorses

The exploration of space has been dominated by the Soviet Union and the United States. More than 98 percent of the rockets launched into space have belonged to these two countries.

The main workhorse of the Soviet space program has been the A-2 rocket. This is still used to launch the manned Soyuz and unmanned Progress spacecraft. Standing 49 m (161 ft.) tall on the launch pad, the A-2 has a main core stage and four boosters: all burn kerosene and liquid oxygen. More powerful Soviet rockets include the Proton (D-1), and Energia rockets, both about 60 m (200 ft.) tall. Energia can lift 100 metric tons into orbit. It is also used to launch the Soviet space shuttle.

▶ An American Delta rocket streaks away from the launch pad at Cape Canaveral, Florida. Its nine solid boosters provide most of the lift-off thrust.

▼ Ariane 3, developed and built by the European Space Agency, on the launch pad in French Guiana. The Ariane rockets were developed to launch satellites for countries without their own rockets.

In the 1980s the United States began to phase out its ordinary, expendable rockets, like the Delta and Titan. It began to switch most of its satellite launchings to the reusable space shuttle. The space shuttle hardware is made of three main units. A winged orbiter carries the crew and payload (cargo). An external tank carries fuel for the orbiter's three liquid-fueled main engines. Two solid-fueled boosters, attached to the external tank, provide additional thrust at lift-off. The space shuttle can carry up to 20 metric tons of payload into low Earth orbit. After a mission the orbiter descends from space and lands on a runway like a normal aircraft.

The space shuttle has not proved as cheap or as reliable as had been hoped. And, following the *Challenger* disaster in 1986, the United States started using expendable rockets once again for routine satellite launchings. The shuttle is now used mainly to support scientific missions. For example, it has been used to carry the space laboratory *Spacelab* and to launch probes to the planets.

The European Space Agency (ESA) has also developed a range of powerful launching rockets, called Ariane. They launch satellites for international customers on a commercial basis. Ariane 3 is a three-stage launch vehicle that can place up to three satellites into geostationary orbit, 35,900 km (22,300 mi.) high. ESA launches the Arianes from its launch site at Kourou in French Guiana.

In the Far East Japan, China, and India have also developed launch rockets. Japan has a thriving space program, and now launches space probes as well as satellites. In 1986 two of its probes encountered Halley's comet and returned much new information.

▼ A Soviet rocket being delivered to the launch pad at the Baikonur Cosmodrome in Central Asia. Hydraulic rams will lift it into the vertical position alongside the launch gantry. The Soviets have more experience at launching rockets than any other country. In the USSR a rocket can be set up and prepared for launch is about two days, compared with about 100 days in the United States.

Units of measurement

Units of measurement

This encyclopedia gives measurements in metric units, which are commonly used in science. Approximate equivalents in traditional American units, sometimes called U.S. customary units, are also given in the text, in parentheses.

Some common metric and U.S. units

Here are some equivalents, accurate to parts per million. For many practical purposes rougher equivalents may be adequate, especially when the quantity being converted from one system to the other is known with an accuracy of just one or two digits. Equivalents marked with an asterisk (*) are exact.

Volume
1 cubic centimeter = 0.0610237 cubic inch
1 cubic meter = 35.3147 cubic feet
1 cubic meter = 1.30795 cubic yards
1 cubic kilometer = 0.239913 cubic mile

1 cubic inch = 16.3871 cubic centimeters
1 cubic foot = 0.0283168 cubic meter
1 cubic yard = 0.764555 cubic meter

Liquid measure
1 milliliter = 0.0338140 fluidounce
1 liter = 1.05669 quarts

1 fluidounce = 29.5735 milliliters
1 quart = 0.946353 liter

Mass and weight
1 gram = 0.0352740 ounce
1 kilogram = 2.20462 pounds
1 metric ton = 1.10231 short tons

1 ounce = 28.3495 grams
1 pound = 0.453592 kilogram
1 short ton = 0.907185 metric ton

Length
1 millimeter = 0.0393701 inch
1 centimeter = 0.393701 inch
1 meter = 3.28084 feet
1 meter = 1.09361 yards
1 kilometer = 0.621371 mile

1 inch = 2.54* centimeters
1 foot = 0.3048* meter
1 yard = 0.9144* meter
1 mile = 1.60934 kilometers

Area
1 square centimeter = 0.155000 square inch
1 square meter = 10.7639 square feet
1 square meter = 1.19599 square yards
1 square kilometer = 0.386102 square mile

1 square inch = 6.4516* square centimeters
1 square foot = 0.0929030 square meter
1 square yard = 0.836127 square meter
1 square mile = 2.58999 square kilometers

1 hectare = 2.47105 acres
1 acre = 0.404686 hectare

Temperature conversions

To convert temperatures in degrees Celsius to temperatures in degrees Fahrenheit, or vice versa, use these formulas:

Celsius Temperature = (Fahrenheit Temperature − 32) × 5/9
Fahrenheit Temperature = (Celsius Temperature × 9/5) + 32

Numbers and abbreviations

Numbers

Scientific measurements sometimes involve extremely large numbers. Scientists often express large numbers in a concise "exponential" form using powers of 10. The number one billion, or 1,000,000,000, if written in this form, would be 10^9; three billion, or 3,000,000,000, would be 3×10^9. The "exponent" 9 tells you that there are nine zeros following the 3. More complicated numbers can be written in this way by using decimals; for example, 3.756×10^9 is the same as 3,756,000,000.

Very small numbers – numbers close to zero – can be written in exponential form with a minus sign on the exponent. For example, one-billionth, which is 1/1,000,000,000 or 0.000000001, would be 10^{-9}. Here, the 9 in the exponent -9 tells you that, in the decimal form of the number, the 1 is in the ninth place to the right of the decimal point. Three-billionths, or 3/1,000,000,000, would be 3×10^{-9}; accordingly, 3.756×10^{-9} would mean 0.000000003756 (or 3.756/1,000,000,000).

Here are the American names of some powers of ten, and how they are written in numerals:

1 million (10^6)	1,000,000
1 billion (10^9)	1,000,000,000
1 trillion (10^{12})	1,000,000,000,000
1 quadrillion (10^{15})	1,000,000,000,000,000
1 quintillion (10^{18})	1,000,000,000,000,000,000
1 sextillion (10^{21})	1,000,000,000,000,000,000,000
1 septillion (10^{24})	1,000,000,000,000,000,000,000,000

Principal abbreviations used in the encyclopedia

°C	degrees Celsius		kg	kilogram
cc	cubic centimeter		l	liter
cm	centimeter		lb.	pound
cu.	cubic		m	meter
d	days		mi.	mile
°F	degrees Fahrenheit		ml	milliliter
fl. oz.	fluidounce		mm	millimeter
fps	feet per second		mph	miles per hour
ft.	foot		mps	miles per second
g	gram		mya	millions of years ago
h	hour		N	north
Hz	hertz		oz.	ounce
in.	inch		qt.	quart
K	kelvin (degree temperature)		s	second
			S	south
			sq.	square
			V	volt
			y	year
			yd.	yard

Glossary

afterburning Technique used to increase the power of jet engines, in which fuel is also burned in a second combustion chamber located behind the turbine.

alternating current Form of electricity supplied to houses and factories, in which the current reverses direction 50 or 60 times per second.

anthracite The best-quality coal. A hard black substance containing at least 86 percent solid carbon. Anthracite gives off little smoke when burned, and is mainly used in household heating.

anticline Dome-shaped formation of underground rock strata in which oil and natural gas often accumulate. About 80 percent of world oil production comes from anticlines.

atom Smallest individual unit of an element. Consists of a nucleus surrounded by one or more orbiting electrons.

beam engine Type of steam engine in which the up-and-down motion of the piston is transferred to a hinged lever or beam.

bimetallic Constructed of two different metals that expand and contract at different rates when heated or cooled.

bituminous coal The most widely distributed form of coal, consisting mainly of solid carbon, but also containing significant quantities of coal tar and gas.

boosters Solid-fueled rocket motors that are often used to provide additional thrust for launching or take-off. Boosters are normally discarded after their fuel has been exhausted.

branch circuit Type of electrical circuit used to distribute electricity to the different parts of a room or floor of a house.

brown coal Low-grade coal that is often soft enough to be crumbled between the fingers, and which is mined in some countries for use in power plants.

bus system Network of thick copper wires and bars used in power stations and substations to carry high-voltage electricity.

camshaft Metal shaft with a series of rounded projections that is used to operate the valves in a four-stroke engine.

carbonization Process by which dead plant material turns to carbon. The term is usually applied to the formation of coal through the action of heat and pressure within the Earth's crust.

carburetor Mechanical device for mixing air and gasoline in the correct proportions for combustion in a gasoline engine.

catalyst Any substance that enables a chemical reaction to take place, or which speeds up a chemical reaction, but which is not itself changed by the reaction.

catalytic converter Device that can be installed in the exhaust system of motor vehicle to reduce the amount of pollution given off in exhaust gases.

cavity magnetron Device that converts electricity into microwaves, and which is used in radar systems and in microwave ovens.

chain reaction Continuous process in which atoms of, for example, uranium split into smaller atoms while releasing large quantities of heat energy.

charcoal Black crumbly substance, consisting almost entirely of carbon, that is made by burning wood in the absence of oxygen.

Christmas tree System of pipes and valves at the top of many oil wells. It is used to regulate the pressure of the oil reaching the surface.

circuit breaker Safety device that breaks an electrical circuit if it becomes overloaded or if there is a sudden power surge.

coal face That part of a coal seam which is being worked in an underground mine.

coal tar Thick, sticky liquid that can be extracted from coal, and which contains many useful chemicals.

coke Coal which has been baked in an oven to remove any gas and coal tar. Coke gives off little smoke when burned and is mainly used by industry and power plants.

combustion chamber That part of a jet or rocket engine in which fuel is burned.

commutator Part of an electric motor or generator (direct current only) that passes current in to or out of the coil by rotating against contacts known as brushes.

compound turbine Type of steam turbine that uses steam at different pressures in a series of cylinders.

condenser Part of a steam engine or steam turbine in which steam is condensed after it has been used to produce power.

containment Thick layer of reinforced concrete that surrounds a nuclear reactor as a safety precaution to prevent the escape of harmful radioactivity.

core Central part of a nuclear reactor in which a chain reaction takes place. Consists typically of uranium fuel rods, a moderator, and control rods.

cracking Process used in the refining of crude oil, in which large hydrocarbon molecules are broken down into smaller ones.

crankshaft That part of a gasoline or diesel engine that transforms the up-and-down movement of the pistons into rotary motion.

Darreius turbine Type of vertical-axis wind turbine with curved blades that are attached to the rotor at each end.

decomposition Process by which dead plant and animal material is broken down by the action of bacteria and other microorganisms.

direct current Form of electricity in which the current flows in one direction only. The main use of direct current is in battery-powered devices (batteries can produce only direct current), but it is also used in some industrial processes.

discharge lamp Type of electric lighting that produces light by exciting the atoms of a gas or vapor. A fluorescent lamp is a special kind of discharge lamp. Ordinary discharge lamps, however, are mainly used for street lighting and advertising because they produce only light of one particular color.

distributor Part of the electrical system of a gasoline engine that supplies high-voltage current to each spark plug in turn.

double-acting Term used to describe steam engines in which every stroke of the piston is driven by steam.

enrichment Process that increases the ability of uranium to sustain a chain reaction, and which therefore makes it a more useful and valuable fuel.

fast breeder reactor Type of reactor that does not require a moderator, and which can be used to create plutonium from uranium fuel.

filament Coil of very fine wire found inside electric light bulbs. Light is produced by the filament glowing white hot.

fission Process by which a large atom splits into two smaller atoms, releasing energy.

flat-plate collector Simple device for collecting solar energy that consists of a water-filled coil behind a glass cover. Flat-plate collectors are normally used for domestic hot water and are usually installed on rooftops.

flywheel Heavy wheel that smooths out the rotary motion produced by reciprocating engines.

fossil Any physical remains of ancient life. Most fossils are just impressions in stone; only under special circumstances is any organic material preserved.

four-stroke engine Type of engine in which power is produced by only one piston stroke in every four.

fractional distillation Basic process of oil refining by which the different hydrocarbons are separated out at different temperatures inside a hollow metal column.

fuel injector Device that injects measured amounts of vaporized fuel into the cylinder of a four-stroke engine. Fuel injectors are found on all diesel engines, but on only some gasoline engines.

fuel rods Long metal cylinders into which uranium fuel is placed before it is loaded into a nuclear reactor.

fuse Safety device in an electrical circuit. Usually a short piece of wire that melts if the circuit becomes overloaded.

gallery Side tunnel in an underground mine, normally dug at right angles to the main tunnel.

gas turbine A turbine engine powered by hot, compressed gas. The name is sometimes applied to jet engines with turbines.

generator Device used to produce electric current by rotating a coil of wire in a magnetic field.

geothermal Relating to heat energy within the Earth's crust caused by natural radioactivity.

graphite Naturally occurring form of pure carbon.

gravity The downward pulling force of the Earth which acts on all objects on the planet's surface or in its atmosphere.

head (of water) Vertical distance traveled by running or falling water.

heat exchanger Device used to transfer heat energy from one fluid to another, but without them coming into direct physical contact.

heating element Coil of resistant wire that gives off useful quantities of heat when an electric current is passed through it.

hydrocarbons Large group of chemical compounds the molecules of which consist almost entirely of hydrogen and carbon atoms. Crude oil contains thousands of different hydrocarbons and is the main source of these substances.

hydroelectric power Electricity that has been produced by the action of moving water against the blades of a turbine.

ignition coil Electrical device that produces the high voltages needed by the spark plugs of a gasoline engine.

impermeable Waterproof or in some other way impenetrable to a liquid or gas. The term is normally applied to rocks that do not permit water to pass through them.

impulse turbine Form of turbine which is rotated by the impact of a fluid, for example steam or water, against a series of cups or buckets mounted around the rim of a wheel.

insulation Any substance which is used to prevent the unwanted transfer of energy. Heat insulation keeps things hot or cold; electrical insulation confines the flow of electricity.

insulator Any substance that is an extremely bad conductor of electricity, for example glass and ceramics. The term also refers to objects made of these substances which are used to insulate electrical wires or cables.

internal-combustion engine Any engine that burns fuel inside the engine itself in order to produce power.

jet engine Any engine that burns a mixture of fuel and compressed air to provide a stream of hot exhaust gas which propels the engine forward.

kerogen Partially decomposed plant and animal material from which oil is formed by heat and pressure within the Earth's crust.

lignite Lowest grade of coal; a brown crumbly substance mainly burned in power plants.

liquefied petroleum gas Collective name for hydrocarbons such as butane and pentane which are gases at normal temperature and pressure, but which are easily liquefied for storage.

longwall Method of underground mining in which coal is taken from a seam along the entire side of a gallery. As the coalface moves back, the roof of the gallery is allowed to collapse.

mantle The stony section of the Earth between the crust and the core.

methane The lightest hydrocarbon and the main constituent of natural gas.

microwaves High-frequency radio waves that can induce heat in objects that they pass through.

migration (of oil and gas) Process by which oil and gas tend to move from one location to another within the Earth's crust.

missile Any nonliving object that travels through the atmosphere before striking the ground or another object. The term is usually applied to rockets used in warfare.

moderator Substance used in the core of a nuclear reactor to slow down, or moderate, neutrons in order that a chain reaction can take place.

molecule Smallest unit of a particular substance consisting of a characteristic arrangement of atoms.

neutron One of the two types of subatomic particle (the other being a proton) that make up the nucleus of most atoms.

nuclear Relating to an atomic nucleus. The term is often used to refer to the production of energy by means of a chain reaction.

nucleus Central part of an atom, in most cases consisting of both protons and neutrons.

oil pan Main lubricating-oil reservoir of a four-stroke engine.

oil shale Type of oil deposit in which crude oil is trapped within a fine-grained rock, and can be removed only with great difficulty.

overburden Rocks and soil covering a coal seam that lies just below the surface.

payload Carrying capacity of a rocket or missile, given in terms of weight.

peat Waterlogged and partially decomposed plant material that is the first stage in the formation of coal. In some parts of the world, dried peat is used as a fuel.

penstocks Water inlets in the face of a dam that channel water to the turbines.

permeable Describes any substance that allows water or some other liquid or gas to pass through it. The term is usually applied to rock.

photovoltaic cell Device that produces small amounts of electrical current when exposed to sunlight. Often called a solar cell.

plutonium Artificial element that does not occur naturally, but is produced inside nuclear reactors.

pollution Unnatural presence of any substance in the environment, usually with harmful effects.

power tower Type of central collection system for producing electricity from solar energy. A large number of mirrors reflect the Sun's rays to the top of the tower, producing enough heat to boil water for turbines.

pumped storage Method of indirectly storing electricity. Surplus electrical power is used to pump water uphill to a reservoir. When additional electricity is required, the water is allowed to flow downhill to drive turbines.

pylon Tall metal framework used to support overhead power lines.

radioactivity The disintegration of atoms of certain substances, including uranium and plutonium, which causes the emission of harmful radiation.

ramjet Simplest form of jet engine. It has no compressor or turbine.

reaction engine Any engine that produces power through the thrust of the exhaust gases, for example jets and rockets.

reaction turbine Type of turbine that is driven by the flow of a fluid through a series of angled blades attached to a shaft. Most turbines used today, whether driven by water, steam, or gas, are reaction turbines.

reactor Basic component of a nuclear power plant. A device for producing heat from a slow, steady chain reaction. A reactor consists of a core, a containment vessel, and a cooling system.

reciprocating engine Any engine that produces power through the action of a piston within a cylinder, for example steam, gasoline, and diesel engines.

rocket 1. Type of engine that burns a mixture of fuel and chemical oxygen, and which can therefore operate outside Earth's atmosphere. 2. Any vehicle propeled through the atmosphere or space by means of a rocket engine; usually applied to vehicles designed to launch objects or people into space.

room-and-pillar Method of underground mining in which coal is removed from a seam in a series of square "rooms," with pillars of coal being left in place to support the roof.

rotary engine Type of engine that uses a specially shaped "piston" to provide direct rotary motion rather than an up-and-down movement.

rotor That part of a mechanical device, for example a turbine or an electric motor, that produces rotary motion.

runner Alternative name for the rotor in some types of water turbine.

sedimentary rock Rock formed from layers of sediment laid down at the bottom of rivers, lakes, and seas.

slip rings Parts of an electric motor or generator (alternating current only) that passes current in or out of the coil.

solar gain Heat energy which buildings receive from sunlight falling on walls, roofs, and through windows.

staging Method of improving the efficiency of steam turbines by allowing the steam to lose its energy in a series of stages.

strip mining Type of surface mining. The coal may be removed from a single large hole that has been dug down to the seam. Alternatively, it may be removed from a series of trenches. The overburden from one trench is used to fill in the previous trench.

substation Installation that reduces the voltage of electrical current by passing it through transformers, before it is distributed.

superheated Describes water or steam that has been heated under pressure to above 100°C (212°F).

switchgear System of electrical switches found in power stations and substations.

traction engine Steam-powered tractor used especially during the late 1800s.

transformer Device used to increase or reduce the voltage of alternating current.

Trombe wall Simple solar heating device that can be built into houses and other buildings. A double wall, with an outer wall of glass separated from a solid inner wall by air.

turbine Any device that produces rotary motion through the action of a gas or liquid on angled blades attached to a shaft.

turboalternators Electrical generators driven by turbines which produce alternating current.

turbofan Type of jet engine fitted to larg aircraft that uses the turbine to power a large fan at the front of the engine. This fan produces most of the engine's thrust.

turbojet Basic jet engine that uses a turbine driven by the exhaust gases to compress air for combustion with kerosene fuel. All the thrust is produced by the exhaust gases.

turboprop Type of jet engine that uses the turbine to power a propeller mounted ahead of the engine. Both the exhaust gases and the propeller produce thrust.

two-stroke engine Simple gasoline engine that produces power with every other stroke of the piston, mainly used in lawnmowers and small motorcycles.

uranium Heaviest naturally occurring element and a principal natural source of nuclear fuel.

volute chamber Spiral casing surrounding the rotor of most types of water turbine.

wellheading Process by which oil can be loaded onto a tanker directly from an undersea oil well.

whipstock Wedge-shaped attachment used for drilling oil wells at an angle.

wind turbine Windmill that is designed to generate electricity.

Index

fluorescent light 62, *62*
fossil fuel 56
four-stroke engine 70, *70*
fractionating column 24, *24*
fractions 24
France 27, 29, 44, 60
Francis, James 54
Francis turbine 54, *54*
fuel injector 72, *72*
fuel rods 28, 30
fuel system 72
fuels 17, 24, 70, 75

G

Gagarin, Yuri 81
gas, natural 16, 17
   deposits 19
   formation of 19
   liquid 22
gas separation plant 24
gas-turbine engine 77
gasoline 24
gasoline engine 70, 71, *71*, 73
generator 49, 52, *53*, 56, *57*, 73
geothermal energy 38, 44
Germany 12, 49, 70, 74, 77
Geysers power plant 4, *44*
Goddard, Robert H. 80, *80*
graphite 30
Great Britain 14, 34
grid 60, *60-61*
gunpowder 80

H

hairdryer *65*
hard coal 10, 13
heat exchanger 30, 52
heating element 62
heating fuel
   gas 17
   oil 16, 24
heavy oil 20, 24
Hero 51
hexane 18
household appliances 59
Hungary 44
hydrocarbons 17, 18, *18*, 24
hydroelectric power 48, 58
hydroelectric power plants 40, *40*
hydrogen 17, 24, 56

I

ignition 70

impulse turbine 50
incongruity *19*
India 52
Industrial Revolution 9, 10, 39
information technology 59
infrared light 62
insulation 34
insulator 60
intercontinental missile 82
internal-combusion engine 66, 70
intrusion *19*
Italy 36, 38, 44

J

Japan 36, 44
jet engine 76, 77
jet plane 76, *76*

K

Kaplan turbine 54, *54*
Kennedy Space Center 83
kerogen 18
kerosene 24
kinetic energy 50

L

light bulb 58, 59, *59*, 62
   invention of 59
   principles of 62
lignite 10, 12, 13
limestone 18
liquefied natural gas (LNG) 22
liquid fuel, in rockets 82
   refrigeration of 82
liquid methane 22
locomotives 66
   diesel 74
longwall mining 14
LPG tanker *22*
lubrication system 72, 73
   in 2-stroke engines 75

M

mechanical excavator 12, *13*
Mediterranean Sea 23, 34
mercury 62
methane 18, 19, 24
microwave oven 63, *63*
mines 12-15, *15*
mining 12-15
   longwall 14

room-and-pillar 14
strip 12
surface 12, 13
underground 14, 15
Mir 84
moderator 29, 30
molecular sieves 24
motor 49, 65, *65*
motorcycle 73, *74*

N

neon 62
neutrons 28, 30
New Zealand 44
Newcomen, Thomas 67
nickel-chromium alloy 59
Nike-Hercules missile *83*
North Sea 16, 20, 23
nuclear energy 28
nuclear fuel 29, 52, 56
nuclear power 26-31
nuclear power plants 28, 52, *52-53*
nuclear reactor 28, 30, 31
   fast-breeder 30, *30*, 31
   gas-cooled 30, *30*

O

offshore oil platform 21
oil 16-25
   crude 16, 18
   formation of 18
   migration of 18
   production of 20
   reserves of 17, *17*
   transport of 22
oil, engine 72
oil field 16, 17
oil pipeline 16
oil pump *73*
oil refinery 24
oil reservoir 72
oil shales 20
oil traps *19*
Otto, N. A. 70
overburden 12
oxygen 11, 82

P

Paraguay 48
Parsons, Charles 49
payload, rocket 81, 85
Pearl Street power station 49
peat 10, 11, *11*

Further reading

Arnold, Guy. *Coal*. New York: Franklin Watts, 1985.

Berger, Melvin. *Switch on, Switch off*. New York: Crowell/Harper, 1989.

Cosner, Sharon. *The Light Bulb: Inventions That Changed Our Lives*. New York: Walker & Company, 1984.

Cross, Wilbur. *Solar Energy*. Chicago: Childrens Press, 1984.

Dineen, Jacqueline. *Coal*. Hillside, N.J.: Enslow Publishers, 1988.

Dineen, Jacqueline. *Energy from Sun, Wind, and Tide*. Hillside, N.J.: Enslow Publishers, 1988.

Energy From the Sun. Morristown, N.J.: Silver Burdett, 1989.

Fleisher, Paul. *Understanding the Vocabulary of the Nuclear Arms Race*. Minneapolis: Dillon Press, 1988.

Gardiner, Brian. *Energy Demands*. New York: Gloucester Press/Watts, 1990.

Gutnik, Martin J. *Electricity: From Faraday To Solar Generators*. New York: Franklin Watts, 1986.

Haines, Gail Kay. *The Great Nuclear Power Debate*. New York: Dodd, Mead, 1985.

Jefferis, David. *Jet Age*. New York: Franklin Watts, 1988.

Macaulay, David. *The Way Things Work*. Boston: Houghton Mifflin Company, 1988.

Mason, John. *Power Station Sun: The Story of Energy*. New York: Facts on File, 1987.

McKee, Robin. *Energy*. New York: Hampsted Press/Watts, 1989.

McKee, Robin. *Solar Power*. New York: Gloucester Press/Watts, 1985.

Pringle, Laurence. *Nuclear Energy: Troubled Past, Uncertain Future*. New York: Macmillan, 1989.

Scott, Elaine. *Oil! Getting it, Shipping It, Selling It*. New York: Warne, Frederick, 1985.

Weiss, Ann E. *The Nuclear Arms Race – Can We Survive It?* Boston: Houghton Mifflin Company, 1983.

Picture Credits

b=bottom, t=top, c=center, l=left, r=right.

ARPL Ann Ronan Picture Library, Somerset. FSP Frank Spooner Pictures, London. MARS Military Archive and Research Services, Braceborough. NASA National Aeronautics and Space Administration, Washington. RF Rex Features, London. RHPL Robert Harding Picture Library, London. SC Spacecharts, Wiltshire. SP Sci Pix, Wiltshire. SPL Science Photo Library, London.

6 SPL/Tom McHugh. 8 Zefa/R. Smith. 10 Spectrum Colour Library. 11b Zefa/G. Kait. 12 RHPL. 13t SP. 13b Colorific/J. Howard. 14 Zefa. 16 SP. 18-19 RHPL. 20l Zefa/D. Cattani. 20r Robin Kerrod. 22 SP. 23 Tony Stone Worldwide. 23 inset FSP. 25 Zefa. 26 Zefa. 27 RF. 27 inset Associated Press. 31 SP. 32 Zefa/W. Mohn. 34 RF. 35 SPL/Tom McHugh. 36 Spacecharts. 37 FSP. 38 Zefa/ E. Christian. 39l, 39r Zefa. 40 Central Electricity Generating Board. 42 SPL/Tom McHugh. 42-43 SPL/Lowell Georgia. 44 Robin Kerrod. 45 Zefa/J. Pfaff. 46 SPL/Robin Scagell. 48 Zefa/Streichan. 49l ARPL. 49r Michael Holford/Royal Institution. 50 Zefa. 52-53 Quadrant Picture Library/*Nuclear Engineering.* 55l G. R. Roberts. 55r Jerry Mason/*New Scientist.* 56bl SPL/Hank Morgan. 56br SPL/U.S. Dept of Energy. 57t Picturepoint Ltd. 57b Zefa/J. Pfaff. 58 Zefa 59l ARPL. 59r Michael Freeman. 61 Picturepoint Ltd. 63t Robin Kerrod. 63b Panasonic. 66 Robin Kerrod. 67l The Ridgeway Archive. 67r ARPL. 68 John Watney Photo Library. 70 All Sport/Vandystadt. 71 Ford Motor Company. 74-5 Spectrum Colour Library/D. J. Heaton. 75b, 76 SP. 77l Quadrant Picture Library. 77r SP. 78 Zefa/E. M. Bordis. 79t SC. 79cr SC/NASA. 80l Robin Kerrod. 80r NASA. 83tl, 83tr SC. 83b MARS/U.S. Army. 84l European Space Agency. 84r SC. 85 SPL/Novosti.